TRANSVAGINAL
COLOR DOPPLER

To Stuart Campbell – who in

1971 taught me the beauty of

diagnostic ultrasound

TRANSVAGINAL COLOR DOPPLER

A Comprehensive Guide to
Transvaginal Color Doppler Sonography
in Obstetrics and Gynecology

Asim Kurjak

The Ultrasound Institute
University of Zagreb
WHO Collaborating Center
Yugoslavia

The Parthenon Publishing Group
International Publishers in Medicine, Science & Technology

Casterton Hall, Carnforth,
Lancs, LA6 2LA, UK

120 Mill Road, Park Ridge,
New Jersey 07656, USA

Published in the UK by
The Parthenon Publishing Group Limited
Casterton Hall, Carnforth,
Lancs, LA6 2LA, England

Published in the USA by
The Parthenon Publishing Group Inc.
120 Mill Road,
Park Ridge,
New Jersey 07656, USA

British Library Cataloguing-in-Publication Data
Transvaginal color doppler.
1. Gynaecology & obstetrics. Diagnosis. transvaginal ultrasonography
I. Kurjak, Asim
618

ISBN 1-85070-331-0

Library of Congress Cataloging-in-Publication Data
Kurjak, Asim
Transvaginal color Doppler : a comprehensive guide to transvaginal color Doppler sonography in obstetrics and gynecology./ Asim Kurjak.
p. cm.
Includes bibliographical references and index.
ISBN 0-85070-331-0 : $125.00
1. Doppler ultrasonography. 2. Generative organs. Female – Ultrasonic imaging. 3. Pelvis – Tumors – Ultrasonic imaging. 4. Ultrasonics in obstetrics. 5. Fetus – Ultrasonic imaging.
I. Title.
[DNLM: 1. Genital diseases. Female – diagnosis. 2. Prenatal Diagnosis. 3. Ultrasonic Diagnosis – methods. WP 141 K947t]
RG527.5.U48K87 1990
618'.047543 – dc20
DNLM/DLC
for Library of Congress 90-14302
CIP

Composition by Ryburn Typesetting Ltd,
Luddendenfoot, Halifax, West Yorkshire
Printed and bound in Great Britain by
Lawrence-Allen (Colour Printers) Ltd,
Weston-super-Mare, Avon

Contents

List of contributors

G. Crvenković
The Ultrasonic Institute
University of Zagreb
P. Miškine 64
41000 Zagreb
Yugoslavia

D.R. Gratton
Diagnostic Ultrasound
Health Sciences Center
Winnipeg
Manitoba
Canada

J.C. Harrington
Diagnostic Ultrasound
Health Sciences Center
Winnipeg
Manitoba
Canada

J. Hustin
The Histopathologic Institute
Loverval
Belgium

E. Jauniaux
Department of Obstetrics and Gynecology
University Hospital Erasmus
Free University of Brussels
Brussels
Belgium

D. Jurković
The Ultrasonic Institute
University of Zagreb
P. Miškine 64
41000 Zagreb
Yugoslavia

G. Kossoff
Ultrasonics Laboratory
Division of Radiophysics, CSIRO
PO Box 5002
West Chatswood
New South Wales 2057
Australia

S. Kupešić-Urek
The Ultrasonic Institute
University of Zagreb
P. Miškine 64
41000 Zagreb
Yugoslavia

A. Kurjak
The Ultrasonic Institute
University of Zagreb
P. Miškine 64
41000 Zagreb
Yugoslavia

C.S. Levi
Department of Radiology
University of Manitoba
Section of Diagnostic Ultrasound
Health Sciences Center
Winnipeg
Manitoba
Canada

E.A. Lyons
Department of Radiology
University of Manitoba
Winnipeg
Manitoba
Canada

M. Miljan
The Ultrasonic Institute
University of Zagreb
P. Miškine 64
41000 Zagreb
Yugoslavia

H. Schulman
Department of Obstetrics and Gynecology
Winthrop University Hospital
Mineola
Long Island
New York 11501
USA

H. Takeuchi
Juntendo University
Tokyo
Japan

I. Žalud
The Ultrasonic Institute
University of Zagreb
P. Miškine 64
41000 Zagreb
Yugoslavia

Foreword

Medicine today is still a strange mixture of empiricism, unexpurgated tradition and advanced technology, of clinical guesswork and scientific measurement.

In the midst of all this change from the human to the more precise, ultrasound is remarkable for its unique ability to glean diagnostic information without discomfort, indignity or known hazard to the patient. It is not surprising, therefore, that from an exciting novelty ultrasound has now become an irreplaceable diagnostic tool. One cannot but wonder how many applications of this relatively simple invention are still hidden in the future. In replacing some of the witchcraft of the past with the technology of the present we cannot fail to recognize the length of the avenues still to be explored in the future.

Transvaginal color Doppler sonography represents a further 'breakthrough' in ultrasound imaging. The technique has radically changed the approach to non-invasive vascular diagnosis. For the first time, it is possible to obtain truly useful information about functional hemodynamic events. Clinical results are particularly encouraging in differentiating tissue characteristics in pelvic tumors and in visualizing blood vessels in tumors. Transvaginal color Doppler is now in its infancy; its potential, however, is exceptional.

Our group was the first to start using this promising new diagnostic modality early in 1987. Three years of experience in clinical investigation, based on over 15 000 patients, are described and illustrated in this book. We believe that the book will be useful for both investigators in the field and those entering it.

Asim Kurjak

1 Basic Physics and Instrumentation of Transvaginal Color Imaging

D.R. Gratton, E.A. Lyons, J.C. Harrington and C.S. Levi

INTRODUCTION

Modern diagnostic ultrasound scanners produce high resolution two-dimensional gray-scale images of underlying anatomy. Color Doppler scanners produce similar images, but, in addition, are able to superimpose a color map over the gray-scale image representing the Doppler shifts received from moving targets within the examined structures. Most often, these Doppler shifts are received from the moving red blood cells within flowing blood. The physics and instrumentation which apply to color Doppler ultrasound are very similar to those which are involved in gray-scale imaging, but do embrace several additional concepts.

BASIC PRINCIPLES

Ultrasound is high-frequency sound energy which travels in the form of a wave. As a wave, it can be described in terms of frequency, wavelength and speed. Each of these parameters affects how the sound interacts within a medium. In addition, the acoustic impedance of the medium also influences this interaction.

In gray-scale imaging, modern transducers have either single or multiple crystals which emit short pulses of ultrasound and which then receive echoes returning from tissue interfaces. These returning echoes are converted to electronic signals which are amplified and electronically processed. The strength of the returning echo is encoded as a digital value, which, in turn, determines the echo brightness (gray-scale) on the display. The time between emission and reception determines the location from which the echo was received and thus its position on the display. When processed at high speed, a real-time two-dimensional gray-scale image is produced.

In color Doppler imaging (CDI), additional pulses are emitted into the region of interest and the returning echoes are processed to detect a change in frequency. This change in frequency, or Doppler shift, indicates that the echo has been reflected from a moving interface. These signals are then further processed and the relevant Doppler information is displayed as a color map which is superimposed over the gray-scale image. In both gray-scale and color imaging, the ultrasound pulses are emitted in the form of a narrow focused beam. The most important characteristics of the beam are those which determine tissue penetration and echo resolution.

Tissue penetration is limited by the attenuation of sound as it propagates through a medium. Within the beam, the frequency of the sound is the most important factor influencing attenuation.

The ability to resolve small interfaces, both along the axis of the beam and perpendicular to it, determines axial and lateral resolution. Here both frequency and the beam width are the most influential factors.

SOUND FREQUENCY

Frequency (f) is the number of complete oscillations a particle performs per second. Ultrasound is sound with a frequency greater than 20 000 cycles per second or 20 kHz (1 kHz = 1000 cycles per second). In diagnostic ultrasound imaging, frequencies in the range 2–10 MHz are employed (1 MHz = 1 000 000 cycles per second). Frequency is determined by the thickness of the ultrasound crystal and is not significantly altered by tissue interactions. Frequency is an important parameter in determining tissue penetration and echo resolution.

WAVELENGTH

Wavelength is the distance between any two identical consecutive points on the waveform. It plays a very important role in determining beam profile and pulse lengths and therefore greatly influences the detail obtainable in an ultrasound image. Wavelength is determined by the frequency of the sound beam and the propagation speed.

PROPAGATION SPEED

Propagation speed is the velocity (v) of the ultrasound pulse within a medium. It is determined by the density and the compressibility of a medium. In general, the greater the stiffness, the higher the propagation speed. The average propagation speed of sound in tissue is 1540 meters per second (m/s) or 1.54 millimeters per microsecond (mm/μs). It varies only slightly in different soft tissues (i.e. muscle, connective tissue, parenchyma). Propagation speed affects acoustic impedance and refraction. It also determines the time–distance relationship of echoes, and therefore their axial location on the display.

WAVELENGTH, FREQUENCY AND VELOCITY

Wavelength, frequency and velocity are related as follows:

$$\text{Wavelength} = \frac{\text{Velocity } (v)}{\text{Frequency } (f)}$$

Wavelength and frequency are inversely related. The higher the frequency, the shorter the wavelength. Because sound velocity is relatively constant in soft tissue, wavelength only changes significantly with a change in frequency. Frequency is determined by the thickness of the piezoelectric crystal in the transducer. Therefore, changing transducers is the only practical means of changing wavelength.

ACOUSTIC IMPEDANCE

Acoustic impedance (Z) is the resistance to sound as it travels through a medium. It is a characteristic of the medium, and is defined as the product of the density of the medium and the sound velocity within the medium. Acoustic impedance is an important factor in the determination of reflection of the sound beam and, therefore, of its echo amplitude. The greater the difference in acoustic impedance between the media forming an interface, the greater the percentage of the sound beam that is reflected back to the transducer, and the greater is the amplitude of the reflected echo.

ATTENUATION

Attenuation is the reduction in amplitude and intensity of sound as it travels through the medium. It is primarily due to absorption (conversion of sound energy to heat), reflection and scattering. Attenuation is measured in decibels (dB), which is a logarithmic unit comparing two sound intensities ($\text{dB} = 10 \log I_1/I_0$; where I_1 is the new intensity level being measured and I_0 is the reference intensity level). A loss of 3 dB in intensity is equal to a 50% decrease in the energy of the sound beam. Attenuation causes deep echoes to be weaker than echoes returning from superficial interfaces. This limits the depth of penetration. The decreased attenuation in fluid-filled structures and the increased attenuation in solid structures permit accurate differentiation of cystic from solid masses (Figures 1.1 and 1.2). Ultrasound imaging systems have circuits which may be manipulated by the operator to compensate for tissue attenuation. The appropriate adjustment of electronic compensation or TGC (time gain compensation) is critical in order to produce images which have balanced echo brightness from near field to far field. Attenuation is very dependent upon frequency. As sound frequency increases, so does attenuation. From a clinical standpoint, the highest frequency transducer which is capable of penetrating an area of interest should be selected for an examination. For transvaginal sonography the usual operating frequency is currently 6.5–7.5 MHz. However, higher frequencies may be used for high resolution images of near full structures. Conversely, lower frequencies such as 5 MHz may be selected to examine larger and deeper structures, or structures that are highly attenuating.

RELEVANT TERMINOLOGY (ATTENUATION)

Not all tissues and materials have the same attenuation characteristics. The following descriptive terms are used to define the attenuation properties of structures within ultrasound images.

Acoustic shadowing is a region of echo dropout due to strong attenuation or refraction. Typically, shadowing is seen behind areas of calcification, stones, leiomyomas, and scar tissue (Figure 1.2).

Acoustic enhancement is the reciprocal of acoustic shadowing. It is a region of increased echo brightness seen behind a structure of low attenuation. It is typically seen behind cysts (Figure 1.3). The presence of acoustic enhancement is a major criterion for the sonographic diagnosis of a fluid-filled structure.

REFLECTION

Reflection is the generation of an echo at an interface between tissues having different acoustic impedances. For clinical purposes, reflection may be described as either specular or non-specular.

Specular reflections are echoes which occur at large, smooth interfaces. Good examples are organ boundaries or vessel walls. Specular reflection is very dependent upon the incident angle. For maximum echo amplitude and detail, the incident sound beam should be perpendicular to the interface (Figure 1.4). At oblique incidence (non-perpendicular) (Figure 1.5), echo amplitude is decreased or the echo may not even return to the transducer. In the latter situation, a refractive edge shadow is displayed. Examples of specular reflectors in the pelvis include the endometrial cavity, and the walls of cysts and vessels. Non-specular reflection or scattering is the production of low amplitude echoes by small interfaces. Scattering is relatively independent of the incident angle; however, it is very dependent on the frequency of sound. As frequency increases, the degree of scattering increases. Scattering occurs within tissue parenchyma, in fluids containing particulate matter, and at relatively rough boundaries. Examples of scatterers in the pelvis include the myometrium, the endometrium and hemorrhagic or infected cysts (Figure 1.6).

RELEVANT TERMINOLOGY (REFLECTION)

The presence or absence of echoes, and the brightness of those echoes within the ultrasound image determine the sonographic characteristics of the examined structures. The following descriptive terms are used to define the reflective properties of objects or tissues.

Anechoic – without internal echoes. It is a typical feature of cysts and vessels. Other words with the same meaning are echolucent, sonolucent or echo free (Figure 1.1).

Hypoechoic – being less echoic than a reference structure. For example, small undegenerated fibroids are typically hypoechoic relative to normal myometrium. Also echopenic or echo poor.

Hyperechoic – being more echoic than a reference structure. For example, endometrium is typically hyperechoic relative to myometrium (Figure 1.2).

Echoic – having internal echoes.

Isoechoic – having the same echo brightness as a reference tissue. For example, if a small intramural leiomyoma is isoechoic to the surrounding myometrium, it is not likely to be detected.

REFRACTION

Refraction is the redirection of the sound wave as it is transmitted across an acoustic interface (Figure 1.5). It occurs at oblique incidence when the tissue velocities are significantly different. Refraction may cause acoustic shadowing at the edge of curved surfaces such as at the edge of cysts or fibroids (Figure 1.7).

AXIAL AND LATERAL RESOLUTION

The detail within an ultrasound image depends on many factors. Several of the most important are the resolution characteristics of the sound beam. The ability to resolve small, closely spaced interfaces depends on their relationship to the beam axis.

Axial or depth resolution is the transducer's ability to resolve two closely spaced interfaces along the axis of the sound beam. It is most dependent on the length of the pulse. Shorter pulses produce better resolution. Pulse length is inversely related to transducer frequency and damping (Figure 1.8). As frequency and damping increase, pulse length decreases and axial resolution improves. The practical means of optimizing axial resolution is to use high frequency transducers and to adjust the power output and receiver gain to acceptable levels, ensuring adequate penetration and echo display. Typical axial resolution is 1 mm.

Lateral or transverse resolution is the transducer's ability to resolve two closely spaced interfaces across the axis of the sound beam. It is most dependent on beam width. Because beam width varies along the beam axis, lateral resolution is not constant with depth. The best lateral resolution is in the focal region. Typical lateral resolution in the focal region is 2–3 mm. Electronic focusing of array transducers extends the focusing range of the transducer and provides a marked improvement in image detail due to the improved lateral resolution.

FOCUSING

Diagnostic ultrasound transducers generate ultrasonic beams which are narrow and highly directional. Transducers are focused in order to minimize beam width and therefore to improve lateral resolution (Figure 1.9). The narrowest part of an ultrasound beam is called the focal zone or focal length. The depth of the focal zone is most dependent on the transducer diameter and frequency. For maximum tissue detail, the area of

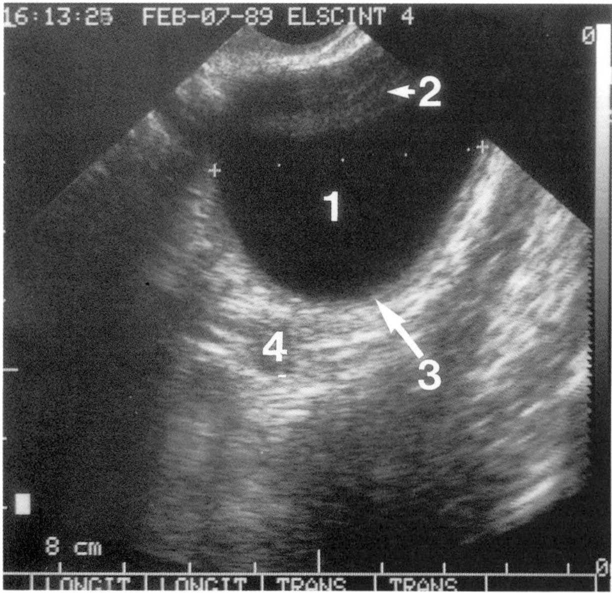

Figure 1.1 Sagittal scan of a simple ovarian cyst performed with a 6.5 MHz transvaginal transducer. The cyst is anechoic (1) except for artifact echoes (2) in the area close to the transducer. Note the well-defined posterior wall (3) and the acoustic enhancement (4) behind the cyst

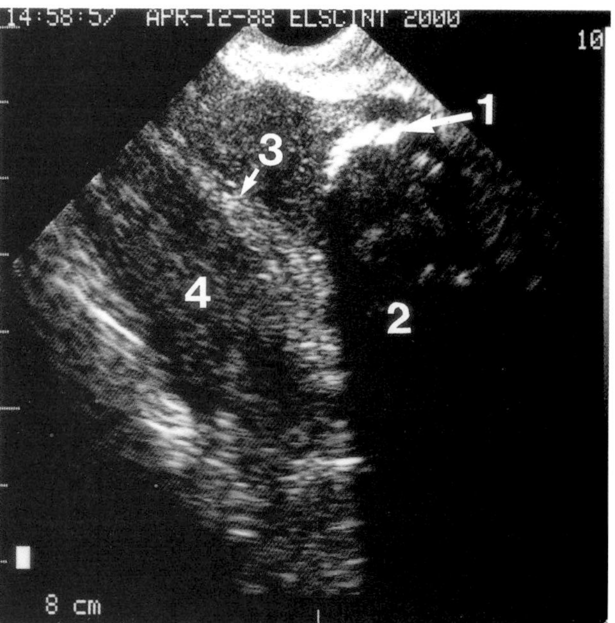

Figure 1.2 Sagittal scan of a uterus showing a bright echo focus (1) in the myometrium consistent with calcification within a leiomyoma. Strong attenuation by the mass produces an acoustic shadow (2). The normal endometrium (3) is noted to be more echoic than the normal myometrium (4)

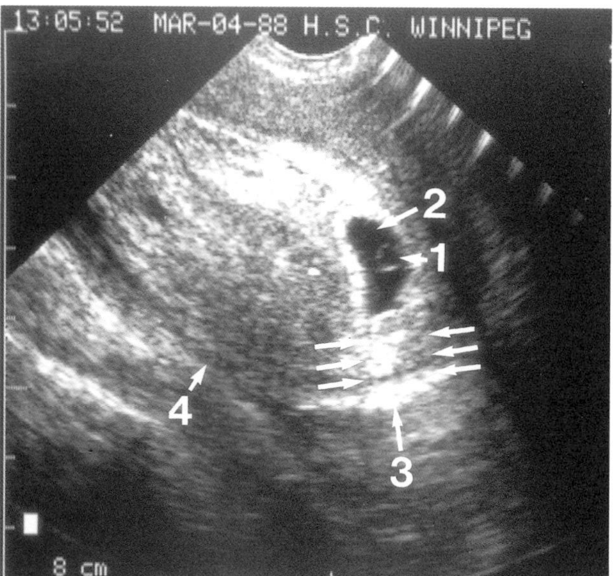

Figure 1.3 Sagittal scan of 5½-week gravid uterus performed with a 6.5 MHz transvaginal probe. Acoustic enhancement is seen behind the sac (arrows). Note that the serosal surface of the uterus deep to the sac (3) is strongly echoic compared to adjacent areas (4). This is due to the incident sound beam being perpendicular to the serosal surface. The increased sound transmission through the gestational sac also adds to the strength of the echoes. A 2 mm yolk sac (1) is seen within a 10 mm gestational sac (2)

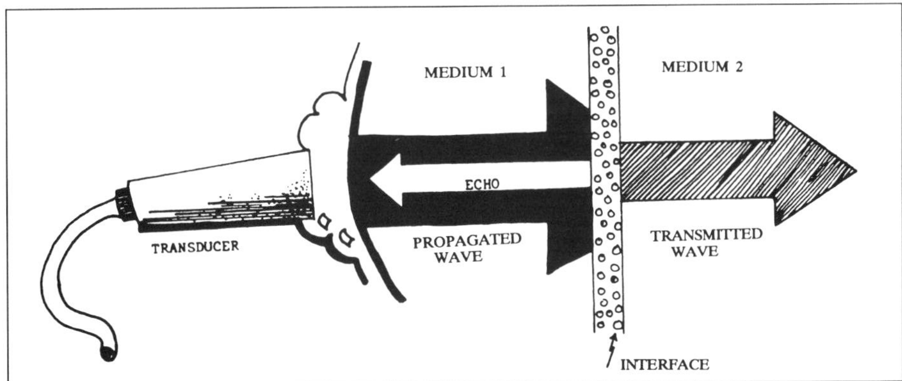

Figure 1.4 Diagram illustrating the concept of specular reflection at perpendicular incidence. Part of the incident wave energy is reflected towards the transducer to produce an echo signal, and part is transmitted into the second medium. At a typical soft tissue interface only a small percentage of sound energy is reflected, usually < 1%. At an air/soft tissue interface, more than 99% of the sound energy is reflected. This makes it impossible to image interfaces which lie deep to air-containing structures

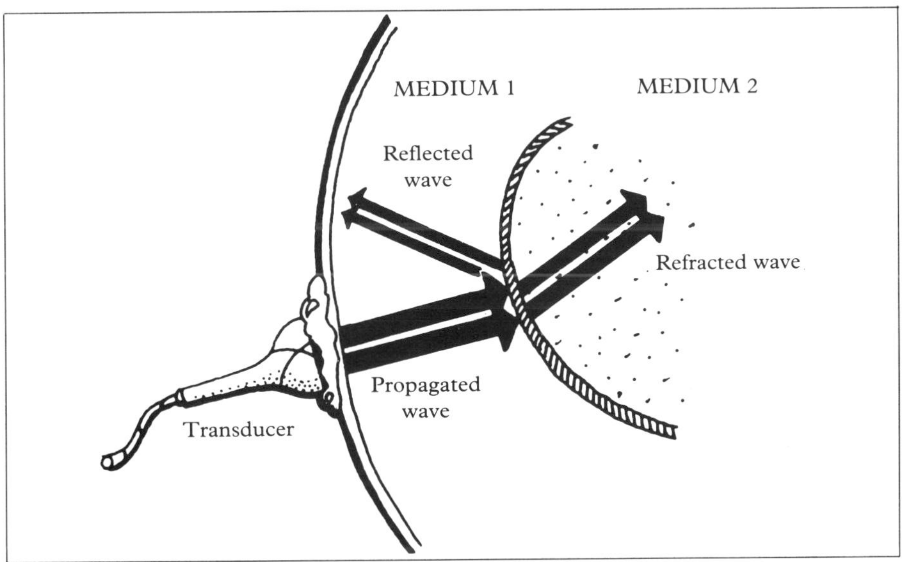

Figure 1.5 Diagram illustrating the concept of oblique reflection and refraction. At oblique incidence, the reflected wave or echo may be small, or may not return to the transducer. Refraction occurs at oblique incidence if the velocities of medium 1 and 2 are significantly different

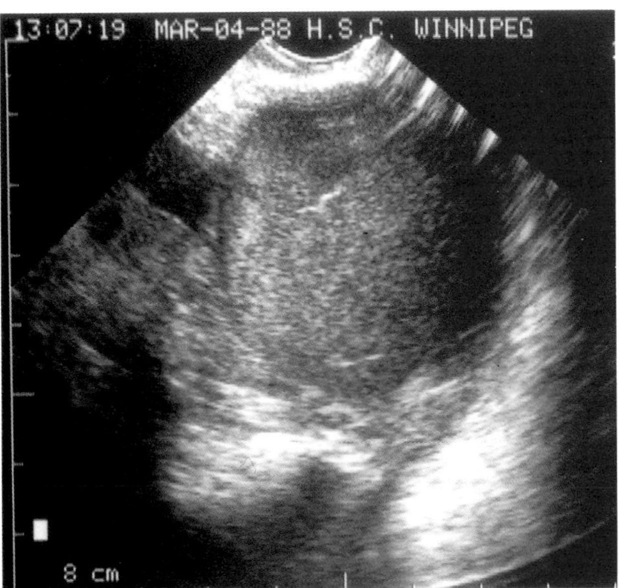

Figure 1.6 Sagittal scan of a hemorrhagic corpus luteal cyst performed with a 6.5 MHz transvaginal probe. The cyst is filled in with low-level internal echoes due to scattering from the blood. Cysts with diffuse internal echoes may appear to be solid structures. However, other criteria for solid masses will be missing. The internal echoes within a cyst are often seen to be mobile on real-time. Gently pushing against the cyst with the transducer often helps to identify fluid movement within the cyst

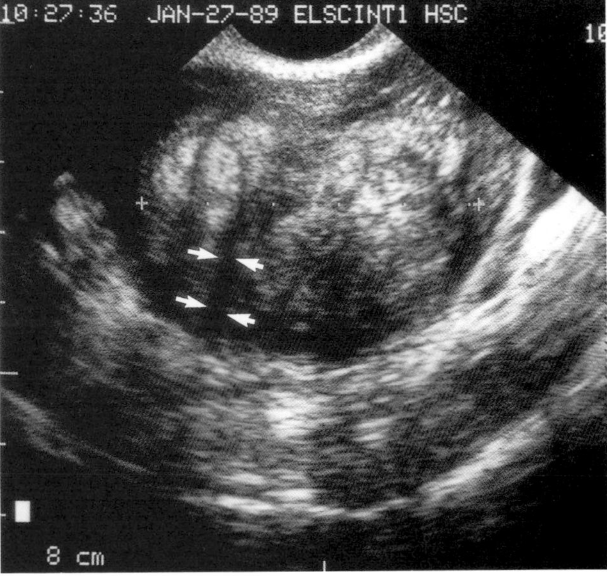

Figure 1.7 Transverse transvaginal scan of the cervix demonstrating a solid, acoustically heterogeneous mass which proved to be a leiomyoma. An edge shadow (arrows) is seen at the edge of a hyperechoic part of the mass. Compared to a cyst, the posterior wall of the mass is irregular and poorly visualized

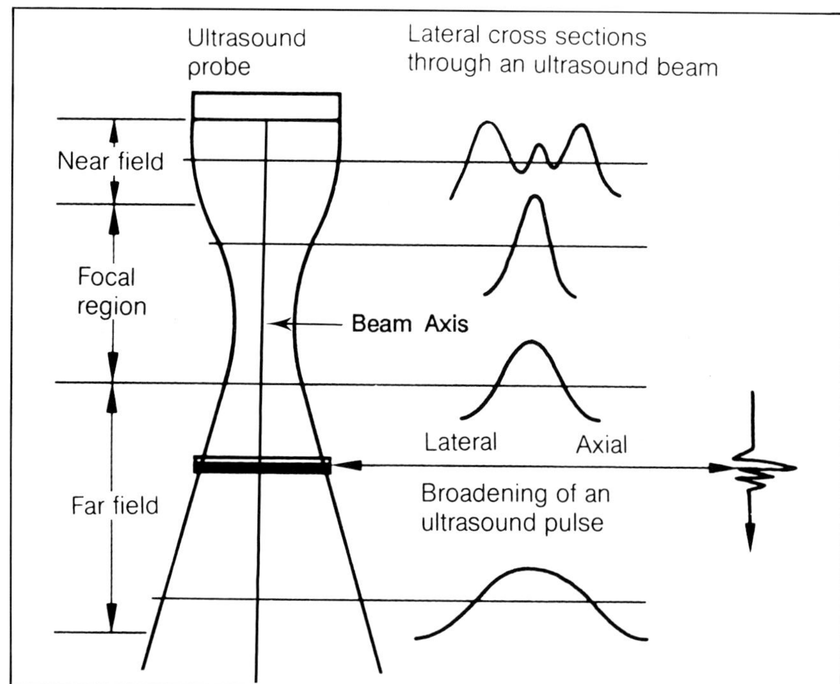

Figure 1.8 Schematic representation of an ultrasound beam divided into three zones – near field, focal region and far field. Lateral resolution varies with depth whereas axial resolution remains relatively constant. Lateral resolution is a function of beam width, and axial resolution is a function of pulse length. Pulse length is very frequency-dependent. (Courtesy of Siemens Electric)

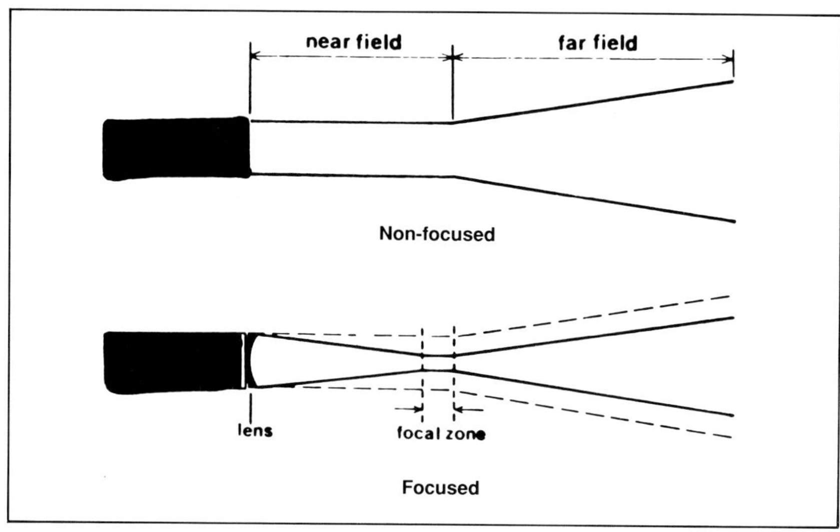

Figure 1.9 Diagram illustrating the concept of both an unfocused and focused ultrasonic beam. The focused beam provides better lateral resolution. The best image detail is seen within the focal zone which is relatively short. Variable focusing is possible with array transducers

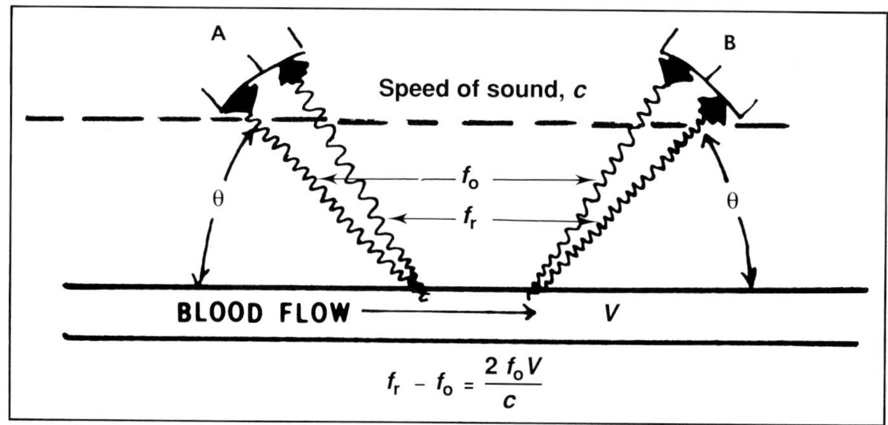

Figure 1.10 Diagram illustrating the concept of Doppler ultrasound for the detection of blood flow. When blood is moving away from the transducer (A), the reflected wave (f_1) has a lower frequency than the transmitted wave (f_0). The converse relationship exists when blood is moving towards the transducer (B). The difference between the transmitted (f_0) and reflected (f_1) frequencies is the Doppler shift frequency. It is dependent on the velocity of sound in the propagating medium (c), the frequency of the transmitted sound (f_0), the velocity of blood (V), and the angle between the transducer and blood flow direction (θ)

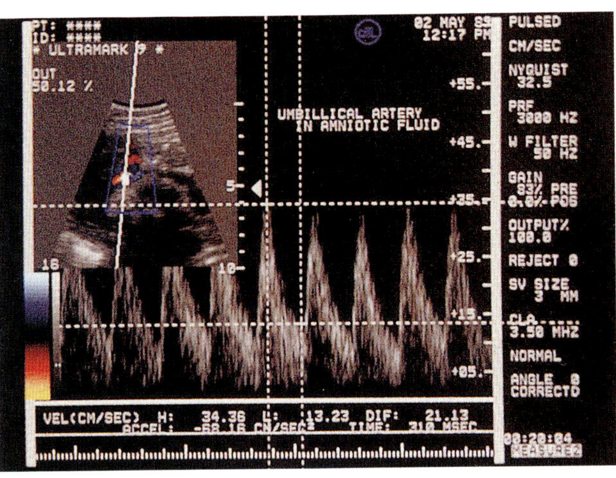

Figure 1.11 Color Doppler image of blood flow in an umbilical cord. The characteristic pulsatile spectral pattern demonstrates that the flow being sampled is within an umbilical artery. (Courtesy of Advanced Technology Laboratories, Bothwell, Washington, USA)

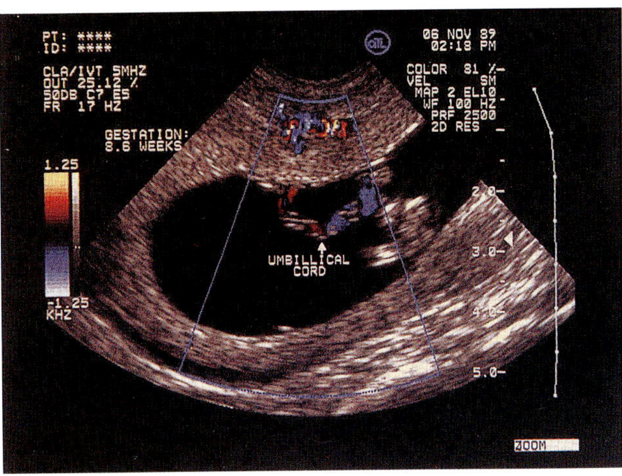

Figure 1.12 Color Doppler image of an 8.6-week pregnancy performed with a 5 MHz endovaginal probe. Blood flow in the umbilical cord is demonstrated with red and blue pixels

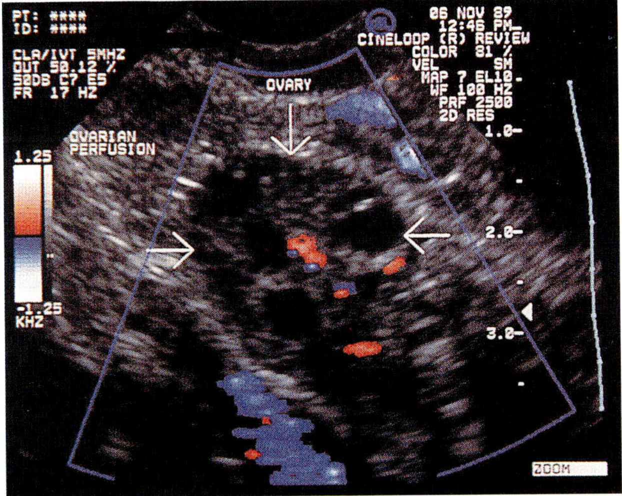

Figure 1.13 Color Doppler image taken with a 5 MHz endovaginal probe. Ovarian perfusion is easily displayed as red and blue color pixels within the ovary. (Courtesy of Advanced Technology Laboratories, Bothwell, Washington, USA)

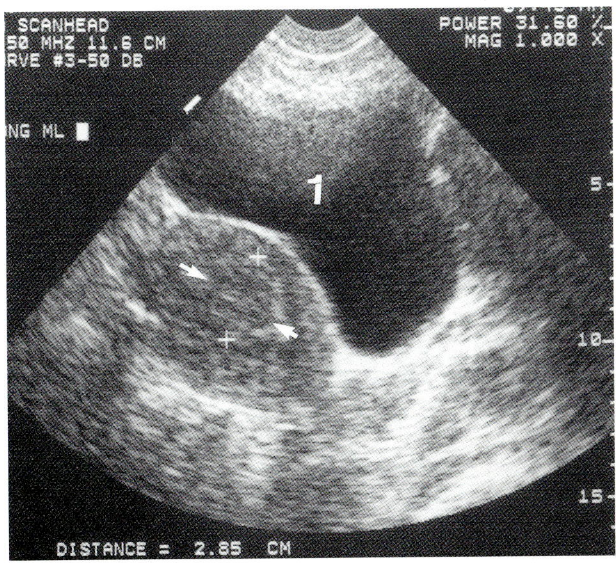

Figure 1.14 Sagittal scan of a non-gravid uterus done with a conventional transabdominal 3.5 MHz transducer. The uterus is displayed posterior to the distended urinary bladder (1). A 2.8 cm poorly defined mass which is slightly more echogenic than the myometrium is seen within the uterus. Inadequate contrast and spatial resolution make it difficult to determine the exact location and nature of the mass

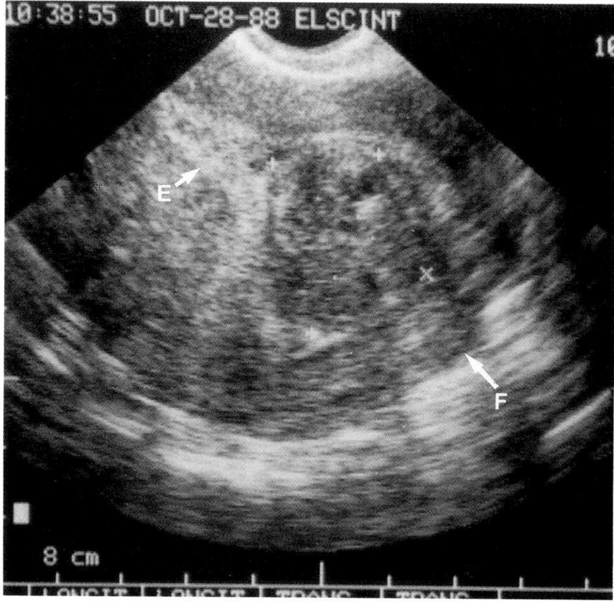

Figure 1.15 Sagittal scan of the same uterus as in Figure 1.14 performed with a 6.5 MHz endovaginal probe. The mass is better defined and appears to be located in the endometrial cavity. The location, heterogeneous echo pattern and relatively poor sound transmission through the mass are most consistent with a submucosal leiomyoma. (E = endometrium; F = uterine fundus)

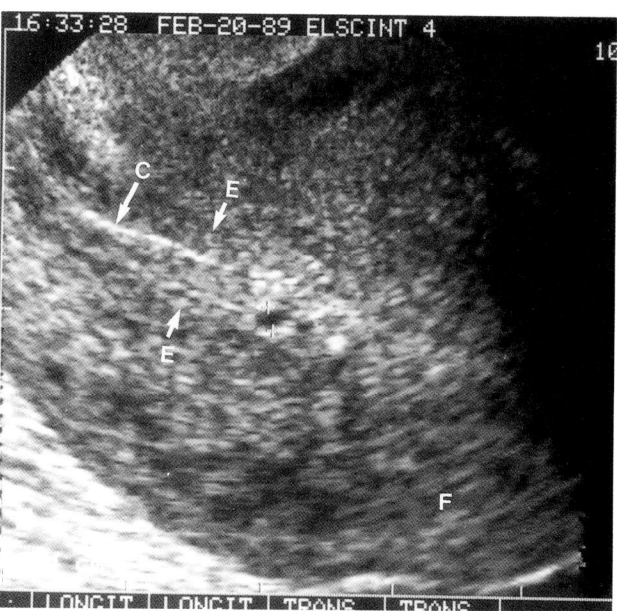

Figure 1.16 Sagittal scan of the uterus done with a 6.5 MHz endovaginal probe. Excellent contrast and spatial resolution permit differentiation of tissues such as the endometrium and myometrium. It also enables visualization of small structures such as this 1.6 mm endometrial cyst of unknown etiology. (C = endometrial cavity; E = endometrium; F = uterine fundus)

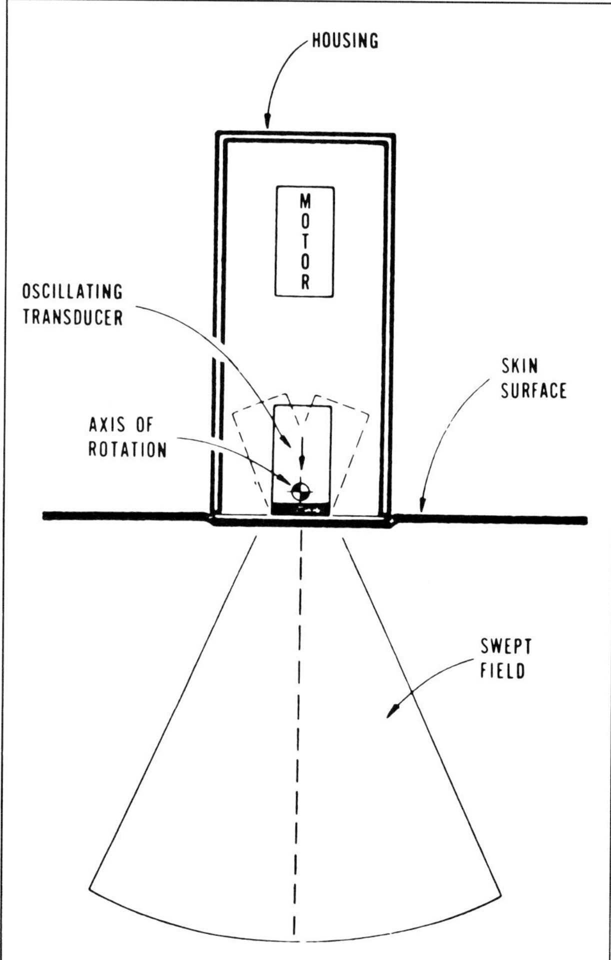

Figure 1.17 Diagram illustrating the concept of a mechanical sector transducer. The transducer is rocked mechanically about an axis of rotation to produce a sector field of view

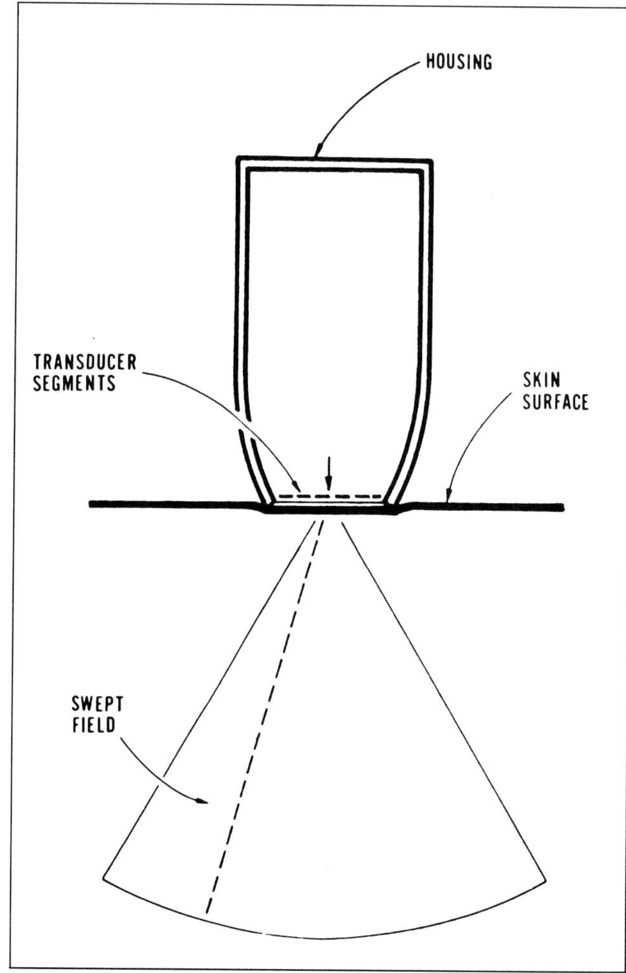

Figure 1.18 Diagram illustrating the concept of a linear phased-array transducer. The sound beam is swept through a sector field view by electronic switching and control of the excitation pulses to the numerous array crystals

interest should be examined within the transducer's focal zone. Structures outside the focal zone (in the near or far field of the beam) are poorly resolved. Single-element transducers have a fixed focal zone. Multiple-element transducers or arrays have a variable focal zone which can be electronically selected by the operator to optimize the lateral resolution within the area of interest.

DOPPLER ULTRASOUND

Doppler ultrasound provides the ability to detect blood flow using the Doppler shift frequency. The Doppler shift frequency is the difference between the transmitted and received ultrasound frequencies (Figure 1.10). Only moving reflectors such as blood cells or vessel walls can produce Doppler shifts. Doppler ultrasound instruments are capable of detecting the Doppler shift signal, and are able to distinguish those signals from echoes returning from static interfaces such as soft tissue. These instruments are capable of detecting the presence or absence of blood flow, and are capable of determining the direction and velocity characteristics of the flow. Spectral display of the Doppler signals provides the ability to display flow waveforms, and to characterize normal and abnormal arterial and venous flow patterns (Figure 1.11). Doppler flow assessment is greatly facilitated by the use of color Doppler imaging which allows the simultaneous overlayed display of anatomic (gray-scale) and flow (color) information (Figures 1.12 and 1.13).

Since a gray-scale format is used to display echoes from tissue, colors are assigned to display the flow data. Two primary colors are used to differentiate flow direction. Typically, blood flow towards the transducer is displayed in red, while blood flowing away from the transducer is displayed in blue. It is important to remember that color selection is completely arbitrary and on most machines can be changed by the operator. A change in color towards white is commonly used to represent an increase in velocity. Green is often added to the display to represent flow variance. It is added proportionately to the display in the presence of disturbed flow. Turbulent flow in color Doppler imaging is typically represented as a mosaic pattern with mixtures of these three colors. Color Doppler imaging presents flow information simultaneously across an entire region of interest, superimposing it on the gray-scale image. The great advantage of color display is that it permits rapid identification of vascular structures, both large and small, and typically results in reduced examination times.

TRANSVAGINAL TRANSDUCERS

Transvaginal probes are able to utilize higher frequency transducers because of the proximity of the organ of interest to the transducer. The advantage of these probes over conventional transabdominal probes is due to the improved resolution produced by the higher frequency sound beam, and to the proximity of the transducer to the area of interest (Figures 1.14–1.16). A disadvantage of the transvaginal probe is that it provides a more limited field of view than that obtained using the transabdominal approach. There are many different designs of transvaginal probe. Most of the types of transvaginal probes include those that utilize a single-element, mechanical oscillating transducer, those using a curved linear array, and those that utilize electronic phased steering of multiple transducer elements known as phased arrays (Figures 1.17 and 1.18). Mechanical transducers are so named because they move the sound beam through the two-dimensional sector arc by means of an electric motor.

Phased-array transducers are generally more complex and more expensive than mechanical systems. Recent advances in solid-state electronics have allowed the development of reliable phased-array systems that provide high quality images. Curved and phased arrays are capable of electronic beam focusing which allows the operator to electronically select the focal zone. Probes capable of color Doppler imaging must use phased-array technology.

In general, the imaging depth of most transvaginal probes is approximately 10 cm, with the focal range varying from 2 to 7 cm, depending upon the type and design of the probes. The image display angle is typically 90–100°, but can be as great as 270°. Different manufacturers provide a variety of transvaginal probes. Some probes employ an end-fired transducer, while others angulate the transducer in relation to the shaft. Still others use a combination of both. Needle guides for biopsy and aspiration are available for most transvaginal probes.

2 Examination Technique

A. Kurjak and I. Žalud

WHY TRANSVAGINAL COLOR DOPPLER SONOGRAPHY?

Conventional transabdominal sonography continues to enjoy the distinction of being the modality of choice to evaluate the abdomen and pelvis sonographically. However, the transabdominal approach in evaluating the female pelvis is being challenged by the enhanced resolution afforded by the recently available transvaginal probe. The detail of the female reproductive tract often seen with transvaginal sonography is exciting to those of us working in this area. It is as if a masterpiece already appreciated for its beauty has undergone restoration, bringing to light exquisite detail and images not before appreciated. After the introduction of transvaginal sonography, this technique was largely used by a few enthusiasts. The use of a high-frequency probe in patients for gynecological examination and for diagnosis of early normal and abnormal pregnancy has been described elsewhere[1–21]. Transvaginal sonography now represents a new tool for gynecological diagnosis and decision making. Its special advantage is that it can reach the uterus and the ovaries at close range. A structural refinement of the surrounding tissue allows one to improve the coordination of individual structures with the different organs. This technique is of special importance in the assessment of uterine and adnexal masses. As well as the demonstration of the endometrium, the time of ovulation can be determined by folliculometry, and an excessive follicular growth can be stopped at the proper time. Transvaginal puncture of follicles is a quicker and easier method than the usual laparoscopic approach. Early diagnosis of normal pregnancy or pregnancy failure can be verified accurately. A cervical insufficiency can be evaluated more effectively by the determination of the opening of the internal orificium of the cervix.

Transvaginal sonography is said to be a revolution in gynecological practice and has come into widespread use (Figure 2.1). Numerous comparative studies have been performed to evaluate the advantages and superiority of transvaginal sonography in comparison with the classical transabdominal approach[22–31]. Moreover, examination of early embryological development becomes more precise, accurate and informative and can be performed approximately 1 week earlier than by the transabdominal route. More information can be obtained using Doppler ultrasound than can be gained by morphological study only. Although the pulsed Doppler technique has been widely used in obstetrics, transvaginal application of this method is a recent development[32,33]. Transvaginal Doppler sonography seems to represent a further 'breakthrough' in ultrasound imaging. This technique has radically changed the approach to non-invasive vascular diagnosis. For the first time we can obtain truly useful information about functional hemodynamic events. It is the unique ability to provide conclusive and specific information on important vascular abnormalities that makes the transvaginal Doppler technique so attractive.

Superb morphological data can be completed by functional information about the blood flow of certain structures (Figures 2.2 and 2.3). Furthermore, more recent technological advances have resulted in transvaginal color Doppler, which yields information non-invasively about blood flow that previously could not be obtained (i.e. vascularization of uterine, ovarian, tumor or embryonal tissue). The system uses pulsed Doppler, which performs flow analysis at multiple points along each scan line of echo data (see Chapter 1). That flow information is then color-coded and displayed on an entire corresponding two-dimensional image. Transvaginal color Doppler is now in its infancy; its potential, however, is tremendous! The color system has distinct advantages over the conventional duplex Doppler system. The position of small and tortuous normal or newly formed vessels in particular and the direction of blood flow can be determined quite rapidly and definitively. Most examiners feel that color Doppler increases examination confidence considerably. Furthermore, color Doppler always indicates direction, velocity and type of blood flow, whereas pulsed Doppler enables quantification of such flow. However, the combination of high quality B-mode images, pulsed Doppler and color Doppler in the

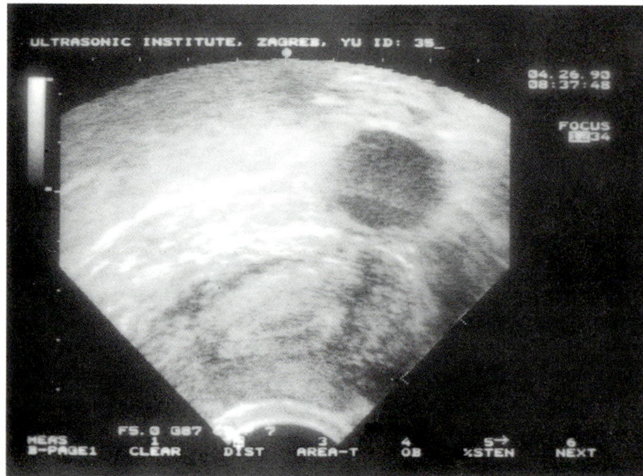

Figure 2.1 A transverse transvaginal scan of the pelvis showing the uterus and the left ovary in the middle of the menstrual cycle. The endometrium is thick and echogenic. The left ovary contains the mature preovulatory follicle

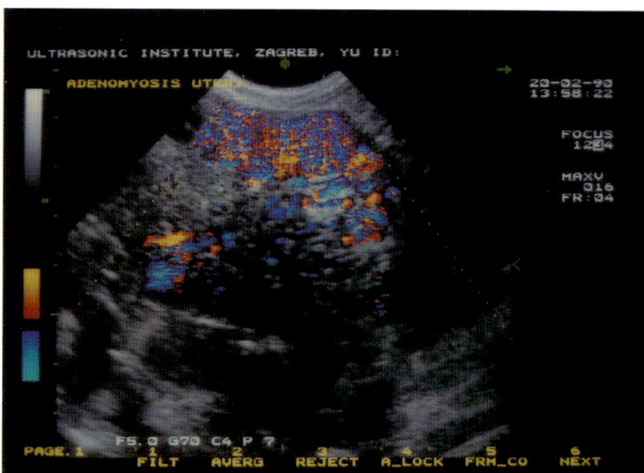

Figure 2.2 Transvaginal color Doppler sonogram showing abundant color flow in small randomly dispersed vessels in the case of uterine adenomyosis. Such small vessels could not be visualized until the advent of color Doppler

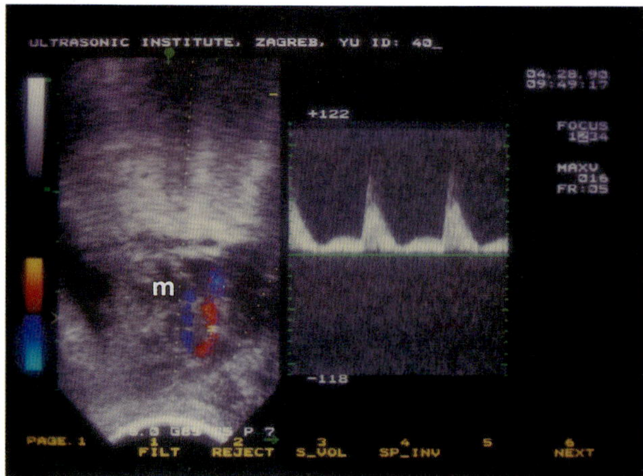

Figure 2.3 A uterine myoma detected by transvaginal ultrasound. Color Doppler visualized blood flow through tumor vessels. Pulsed Doppler analysis (right) indicates high-velocity and high-resistance blood flow. This is a typical finding of normal uterine vessels. m = myoma

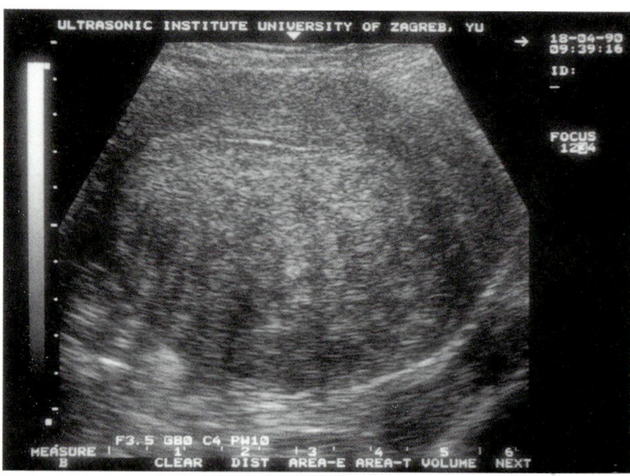

Figure 2.4 A transverse section of an anteverted uterus. The endometrium is thin. The discrete cavity line is normal in appearance. A slightly enlarged uterus and myometrial unhomogenicity were described. The patient was at the beginning of her cycle

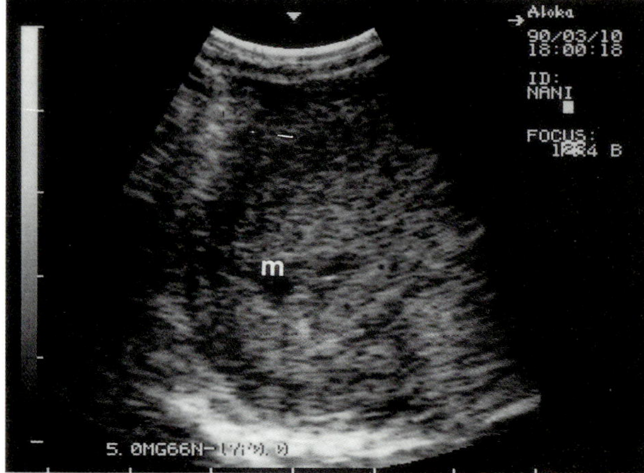

Figure 2.5 Another example of transverse scan of the uterus. The normal uterine texture is destroyed by an intramural myoma. The transonic area indicated secondary changes of the myoma. m = myoma

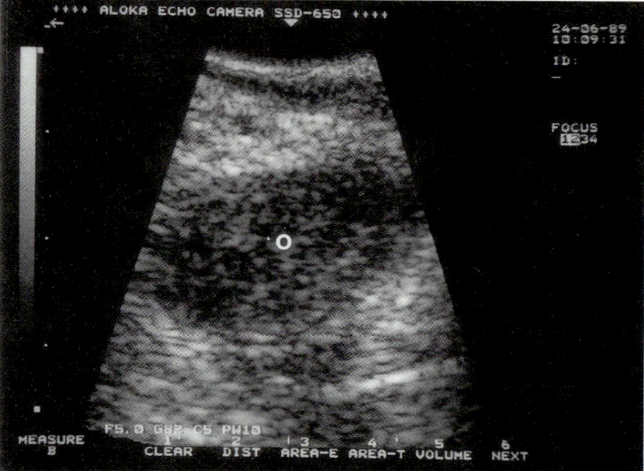

Figure 2.6 The oblique scan through the right ovary exhibits a normal shape and texture. o = ovary

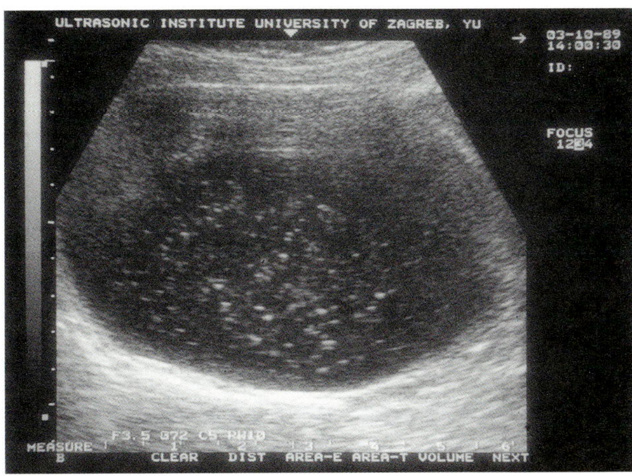

Figure 2.7 A large adnexal cyst with pronounced internal echoes. Such a finding is usually diagnosed as endometrioma or a tubo-ovarian abscess. However, the morphological data are not typical enough to differentiate benign and malignant adnexal masses

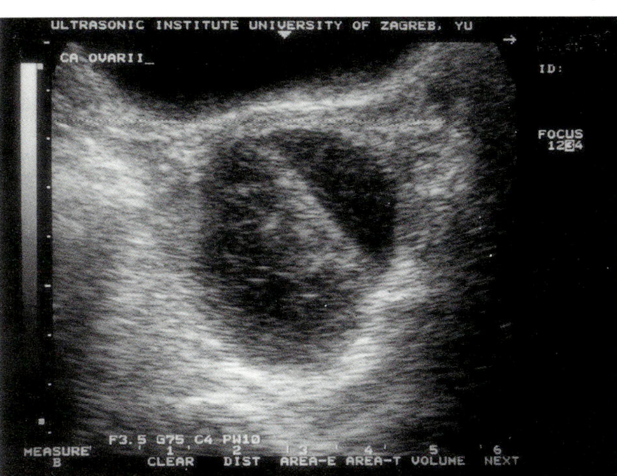

Figure 2.8 A complex, predominantly solid adnexal mass detected by transabdominal ultrasound. The bizarre appearance suggested a malignant tumor

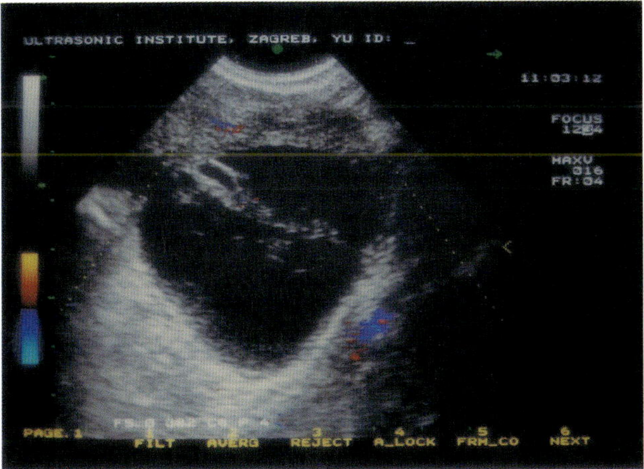

Figure 2.9 The same patient. Transvaginal ultrasound offered more morphological information. Color Doppler presented poor vascularization at the tumor periphery

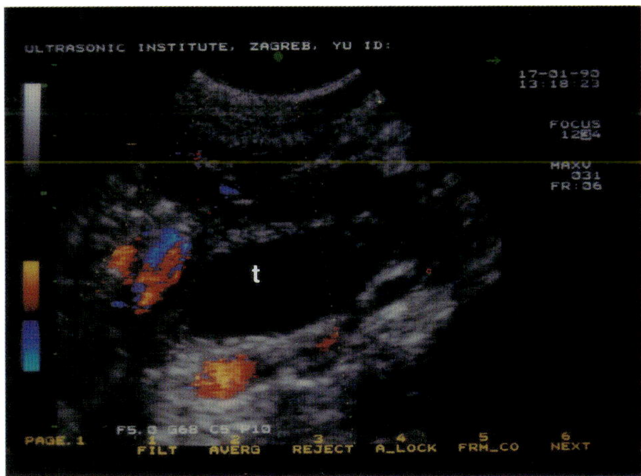

Figure 2.10 Prominent color flow indicated good vascular tumor supply. Large vessels surrounded the adnexal mass. t = adnexal tumor

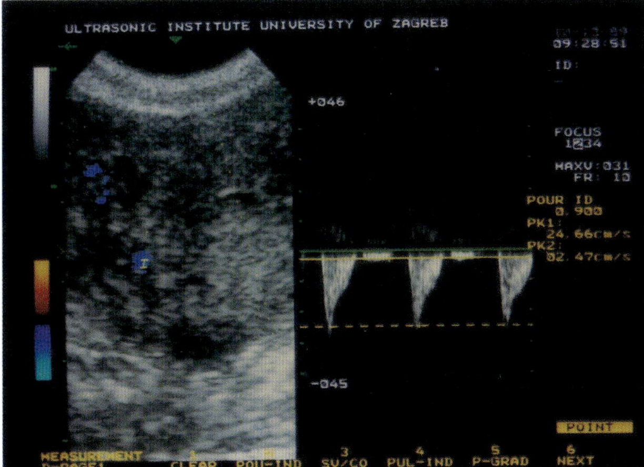

Figure 2.11 A transverse scan of the enlarged, unhomogeneous uterus. A sample volume in the lateral part of the uterus showed a high-impedance blood flow with a small hump during diastole. The waveform (right) is of a high-impedance pattern (RI = 0.900) and is consistent with one of the terminal branches of the uterine artery

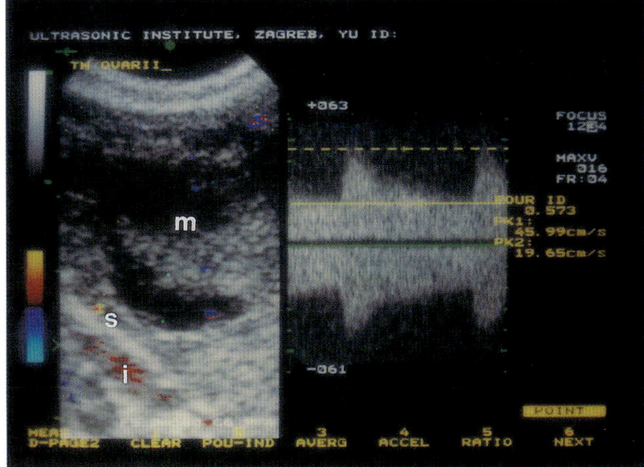

Figure 2.12 Pulsed Doppler signals (right) were obtained from color-coded areas. Moderate velocity, increased diastolic flow and decreased RI (0.573) were present. Morphological study supported the diagnosis of a malignant tumor and Doppler analysis was typical for benign lesions. The benign nature of the adnexal tumor was confirmed by surgery and histopathology. s = pulsed Doppler sample volume; m = adnexal mass; i = internal iliac vessel

same vaginal probe produces a superb simultaneous visualization of morphological and blood flow information from the female pelvic circulation. Basically, each vessel visualized by color Doppler can be explored by pulsed Doppler. This is a major advantage when flow is to be studied in specific vessels. Flow can be assessed without interference from other vessels lying distally or proximally in the same axis of investigation and having a different circulatory pattern.

This type of coupled transvaginal color and pulsed Doppler exploration has been recently developed and we have shown its application by studying structures that have not been previously described. Transvaginal color Doppler saves time in pulsed Doppler examination and increases accuracy of measurement. The most important advantage of this new technique is the display of blood flow across the whole scanning plane of the pelvis. The pulsed Doppler sample volume can be placed very precisely on the color flow of interest and spectral waveform analysis can be easily performed. We think that the possibility of adding color Doppler capabilities to superb transvaginal imaging will become important when making a selection from the devices currently available on the market.

PATIENT PREPARATION

A brief explanation is required to ensure acceptance of the new scanning technique and full cooperation by all patients. By comparing the insertion of the probe with the insertion of a tampon, speculum examination, or pelvic bimanual examination, the patient's fears are quickly dispelled. Furthermore, the whole procedure can be done in a relaxed atmosphere without any hurry or the inconvenience to the patient of a full bladder.

All women should have a completely empty urinary bladder, so both patient and examiner do not have to wait for the bladder to fill slowly in order to receive satisfactory pelvic images. This might be important in cases of emergency such as ectopic pregnancy, acute pelvic inflammatory disease, and other pathologies where the patient is potentially a subject for surgery and, consequently, may not drink in order to fill the bladder.

Initially, we thought that patients might reject the procedure because they would perceive it to be painful. However, after an explanation, as mentioned earlier, patient acceptance is almost universal. The relative discomfort does not reach a level such that the procedure has to be interrupted before its completion.

SCANNING TECHNIQUE

A regular gynecological examination table with heel support is adequate for vaginal scanning, but a flat examination table can also be used. The patient is scanned in the lithotomy position with a slight reverse Trendelenburg tilt to localize free fluid in the pouch of Douglas. If present, this fluid creates tissue–fluid interfaces, which improve the outline of pelvic structures.

Normal menstruating women are best scanned during days 3–10 of the cycle to exclude changes in intraovarian blood flow that are known to occur during the formation of the corpus luteum[34, 35]. The 5–10 MHz transvaginal transducer produces a sector angle of 90–320° to allow a sufficiently satisfactory view of the pelvic organs. Before it is used, the probe should be covered with a coupling gel and inserted into a condom, and then gently inserted into the vagina. Pelvic structures may be brought closer to the acoustic focal range by placing the operator's second hand on the patient's abdomen, as in bimanual examination. The scanning is carried out systematically, starting with the uterus as a landmark (Figures 2.4 and 2.5). The ovaries are scanned next, then the Fallopian tubes, and finally the cul-de-sac. Various planes and depths are reached by the operator by tilting and angling the tip of the probe, thus introducing various structures into the acoustic focal range. The morphology and size of the pelvic structures should be assessed (Figures 2.6–2.8).

After visualization of the pelvic anatomy by B-mode, color Doppler sonography is used to locate blood flow in normal or newly formed pelvic vessels (Figures 2.9 and 2.10). Subsequently, structures of particular interest are examined for prominent areas of vascularization, probably reflecting neovascularization. These vessels usually appear as continuously fluctuating color rather than the pulsative color seen with normal arteries. The color flow of interest so obtained may be explored with Doppler sample volume until the typical spectral waveform is seen (Figures 2.11–2.14). The angle of the transducer should be moved to obtain the maximum amplitude and clarity of the waveform.

Quantification of color flow can be performed by pulsed Doppler waveform analysis. The peak-systolic (A) and end-diastolic (B) Doppler shift frequency can be recorded and the A/B ratio, the Pourcelot resistance index or the pulsatility index may be calculated[36]. The Pourcelot resistance index (RI) is a useful method of expressing blood flow impedance distal to the point of sampling[37]. Each separate parameter is angle-dependent, but once it is correctly related, the resistance index becomes independent of the angle between the investigated vessels and the emitted ultrasound beam. It is believed that an

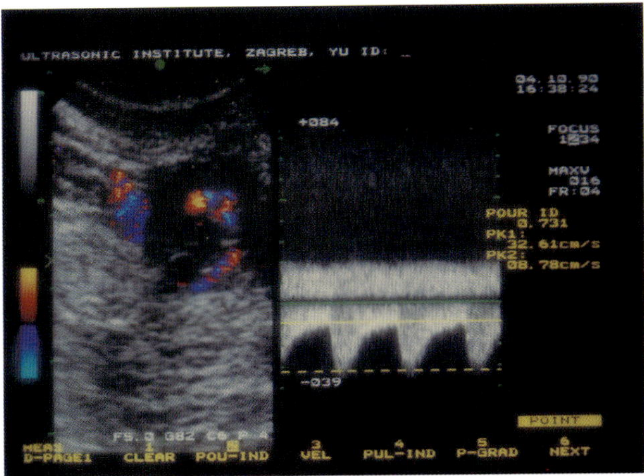

Figure 2.13 An adnexal mass represented by abundant color flow. Pulsed Doppler analysis (right) showed high-resistance flow indicating the benign nature of the tumor. Normal pre-existing vessels can be visualized

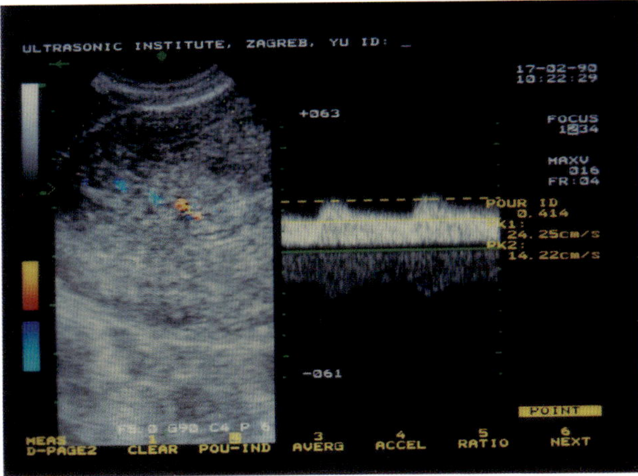

Figure 2.14 A solid adnexal mass. Moderate color flow was noted in the central part of the tumor. The waveform pattern (right) shows low-velocity and very low-impedance blood flow (RI = 0.414). A malignant ovarian tumor was proven

increased value of the resistance index results from increased peripheral vascular resistance[38]. On each measurement, five separate cardiac cycles have to be recorded and the mean value of the resistance index calculated.

The mean duration of examination is usually no longer than 15 minutes in experienced hands. The spatial-peak temporal average intensity should not exceed 100 mW/cm^2, which is the highest permitted limit of insonation energy recommended by the Food and Drug Administration of America for use in fetal medicine.

ADVANTAGES AND DISADVANTAGES OF TRANSVAGINAL COLOR DOPPLER

The advantages of transvaginal color Doppler scanning are prominent and include:

(1) Excellent assessment of the pelvic structures is achieved.

(2) The closeness of the probe to the pelvic organs produces high-resolution sonograms even when there is an abundantly gas-filled bowel, adhesions, or obesity.

(3) Rapid performance of blood flow visualization of the entire insonated area may be carried out.

(4) Small, randomly dispersed and newly formed tumor vessels (neovascularization) could not be visualized prior to the use of color Doppler.

(5) The color signal allows rapid and very precise placement of the sample volume for Doppler waveform analysis.

(6) Normal and pathological blood flow can be distinguished.

(7) Even early embryonal blood flow can be studied.

(8) The combination of morphological and functional (blood flow) studies makes ultrasound diagnosis more accurate.

(9) It has good patient acceptance, especially without the discomfort of a full bladder.

(10) The result of the examination is available immediately.

There are only a few disadvantages of the transvaginal color Doppler sonography:

(1) The color Doppler equipment is expensive.

(2) The field of view is limited.

(3) Structures out of the focal range (i.e. highly placed ovaries) cannot be imaged.

(4) In the second and third trimesters of pregnancy, only a presenting part of the fetus can be examined.

REFERENCES

1. Timor-Tritsch, I.A., Bar-Yam, Y., Elgali, S. and Rottem, S. (1988). The technique of transvaginal sonography with the use of a 6.5 MHz probe. *Am. J. Obstet. Gynecol.*, **158**, 1019
2. Rodriguez, M.H., Platt, L.D., Medearis, A.L., Lacarra, M. and Lobo, R.A. (1988). The use of transvaginal sonography for evaluation of postmenopausal ovarian size and morphology. *Am. J. Obstet. Gynecol.*, **159**, 810
3. Schwimer, S., Rotham, C., Lebovic, J. and Oye, D. (1984). Transvaginal pelvic ultrasonography. *J. Ultrasound Med.*, **3**, 381
4. Meldrum, D., Cherkowski, R., Steingold, K. and Randle, D. (1984). Transvaginal ultrasound scanning of ovarian follicles. *Fertil. Steril.*, **42**, 803

5. Timor-Tritsch, I.E. and Rottem, S. (1988). *Transvaginal Sonography*. (New York: Elsevier)
6. Tessler, F.N., Perrela, R.R., Fleischer, A.C. and Grant, E.G. (1989). Transvaginal sonographic diagnosis of dilated Fallopian tubes. *Am. J. Roentgenol.*, **153**, 523
7. Dodson, M.G. and Deter, R.L. (1990). Definition of anatomical planes for use in transvaginal sonography. *J. Clin. Ultrasound*, **18**, 239
8. Itskowitz, J., Boldes, R., Levron, J. and Thaler, I. (1990). Transvaginal ultrasonography in the diagnosis and treatment of infertility. *J. Clin. Ultrasound*, **18**, 248
9. Lewit, N., Thaler, I. and Rottem, S. (1990). The uterus: a new look with transvaginal sonography. *J. Clin. Ultrasound*, **18**, 331
10. Fleischer, A.C., Gordon, A.N., Entman, S.S. and Kepple, D.M. (1990). Transvaginal scanning of the endometrium. *J. Clin. Ultrasound*, **18**, 337
11. Rottem, S., Lewit, N., Thaler, I., Yoffe, N., Bronstein, M., Manor, D. and Brandes, J.M. (1990). Classification of ovarian lesions by high-frequency transvaginal sonography. *J. Clin. Ultrasound*, **18**, 359
12. Steinkampf, M.P. (1988). Transvaginal sonography. *J. Reprod. Med.*, **33**, 931
13. Modica, M.M. and Timor-Tritsch, I.E. (1988). Transvaginal sonography provides a sharper view into the pelvis. *J. Obstet. Gynecol. Neonatal. Nurs.*, **17**, 89
14. Popp, L.W., Lueken, R.P., Muller-Holve, W. and Lindemann, H.J. (1983). Gynecologic endosonography: initial experience. *Ultraschall Med.*, **4**, 92
15. Timor-Tritsch, I.E. and Rottem, S. (1988). Review of transvaginal ultrasonography: a description with clinical application. *Ultrasound Q.*, **6**, 1
16. Bernaschek, G. and Deutinger, J. (1989). Endosonography in obstetrics and gynecology: the importance of standardized image display. *Obstet. Gynecol.*, **74**, 917
17. Bree, R.L., Edwards, M., Bohm-Velez, M., Beyler, S., Roberts, J. and Mendelson, E.B. (1989). Transvaginal sonography in the evaluation of normal early pregnancy: correlation with HCG level. *Am. J. Roentgenol.*, **153**, 75
18. Timor-Tritsch, I.E., Farine, D. and Rosen, M.G. (1988). A close look at early embryonic development with the high-frequency transvaginal transducer. *Am. J. Obstet. Gynecol.*, **159**, 676
19. Goldstein, S.R., Snyder, J.R., Watson, C. and Danon, M. (1988). Very early pregnancy detection with endovaginal ultrasound. *Obstet. Gynecol.*, **72**, 200
20. Goldstein, S.R. (1990). Early detection of pathologic pregnancy by transvaginal sonography. *J. Clin. Ultrasound*, **18**, 262
21. Nyberg, D.A., Mack, L.A. and Laing, F.C. (1988). Early pregnancy complications: endovaginal sonographic findings correlated with human chorionic gonadotropin levels. *Radiology*, **167**, 619
22. Fleisher, A.C., Gordon, A.N. and Entman, S.S. (1989). Transabdominal and transvaginal sonography of pelvic masses. *Ultrasound Med. Biol.*, **6**, 529
23. Mendelson, E.B., Bohm-Velez, M., Joseph, N. and Neiman, H.L. (1988). Gynecologic imaging: comparison of transabdominal and transvaginal sonography. *Radiology*, **166**, 32
24. Bartrum, J.R. (1989). Transabdominal and transvaginal sonography in early pregnancy. *Am. J. Roentgenol.*, **152**, 1132
25. Jain, K.A., Hamper, U.M. and Sanders, R.C. (1988). Comparison of transvaginal and transabdominal sonography in the detection of early pregnancy and its complications. *Am. J. Roentgenol.*, **151**, 1139
26. Leibman, A.J., Kruse, B. and McSweeney, M.B. (1988). Transvaginal sonography: comparison with transabdominal sonography in the diagnosis of pelvic masses. *Am. J. Roentgenol.*, **151**, 89
27. Brown, J.E., Thieme, G.A., Shah, D.M., Fleischer, A.C. and Boehm, F.H. (1986). Transabdominal and transvaginal endosonography: evaluation of the cervix and lower uterine segment in pregnancy. *Am. J. Obstet. Gynecol.*, **155**, 721
28. Tessler, F.N., Schiller, V.L., Perrella, R.R., Sutherland, M.L. and Grant, E.G. (1989). Transabdominal versus endovaginal pelvic sonography: prospective study. *Radiology*, **170**, 553
29. Coleman, B.G., Arger, P.H., Grumbach, K., Menard, M.K., Mintz, M.C., Allen, K.S., Arenson, R.L. and Lamon, K.A. (1988). Transvaginal and transabdominal sonography: prospective comparison. *Radiology*, **168**, 639
30. Pennell, R.G., Baltarowich, O.H., Kurtz, A.B., Vilaro, M.M., Rifkin, M.D., Needelman, L., Mitchell, D.G., Mervis, S.A. and Goldberg, B.B. (1987). Complicated first-trimester pregnancies: evaluation with endovaginal ultrasound versus transabdominal technique. *Radiology*, **165**, 79
31. Granberg, S. and Wikland, M. (1987). Comparison between endovaginal and transabdominal transducers for measuring ovarian volume. *J. Ultrasound Med.*, **6**, 649
32. Schaaps, J.P. and Soyeur, D. (1989). Pulsed Doppler on a vaginal probe; necessity, convenience or luxury? *J. Ultrasound Med.*, **8**, 315
33. Thaler, I., Manor, D., Rottem, S., Timor-Tritsch, I.E., Brandes, J.M. and Itskowitz, J. (1990). Hemodynamic evaluation of the female pelvic vessels using a high-frequency transvaginal image-directed Doppler system. *J. Clin. Ultrasound*, **18**, 364
34. Baber, R.J., McSweeney, M.B., Gill, R.W., Porter, R.N., Picker, R.H., Warren, P.S., Kossoff, G. and Saunders, D.M. (1988). Transvaginal pulsed Doppler ultrasound assessment of blood flow to the corpus luteum in IVF patients following embryo transfer. *Br. J. Obstet. Gynaecol.*, **95**, 1226
35. Žalud, I. and Kurjak, A. (1990). The assessment of luteal blood flow in pregnant and non-pregnant women by transvaginal color Doppler. *J. Perinat. Med.*, **18**, 215
36. Thompson, R.S., Trudinger, B.J. and Cook, C.M. (1988). Doppler ultrasound waveform indices: A/B ratio, pulsatility index and Pourcelot ratio. *Br. J. Obstet. Gynaecol.*, **95**, 581
37. Pourcelot, L. (1974). Applications clinique de l'examen Doppler transcutane. In Peronneau, P. (ed.) *Velocimetre ultrasonore Doppler*, Vol. 34, pp. 213–40. (Paris: Inserm)
38. Kurjak, A., Alfirevic, Z. and Miljan, M. (1988). Conventional and color Doppler in the assessment of fetal and maternal circulation. *Ultrasound Med. Biol.*, **14**, 337

3 Normal Pelvic Blood Flow

A. Kurjak and I. Žalud

The observation of blood flow in the main pelvic vessels is not new, but only lately has it gained popularity. Transvaginal color Doppler is a new technique giving information that relates to the physiological state of the pelvic circulation. To gain an understanding of the information gathered in this way, it must be considered in association with the anatomical structure of the organ under investigation. It is suggested that, despite their common origin, each vessel in the pelvis has a different 'Doppler signature' that can be used to aid its identification[1, 2]. Of considerable scientific and clinical interest is the reason for these differences. It is clear that the perfusion requirements of each pelvic organ determine the state of its vascular bed, which in turn affects the flow characteristics of its associated vessel.

THE MAIN PELVIC VESSELS

Blood flow in the main pelvic vessels can be easily visualized and recognized by transvaginal color Doppler (Table 3.1). The artery and vein are distinguished according to the pulsation and brightness of color flow. The most prominent color and pulsation are associated with arterial flow.

The external iliac vessels are situated laterally to the ovary, presenting abundant color because of the high velocity of blood flow (Figures 3.1 and 3.2). *The internal iliac vessels* can be visualized most easily. They can be observed in the side wall of the pelvis, often lying deep and close to the ovary (Figures 3.3–3.5). Both the internal and external iliac arteries produce prominent and pulsating color flow, high velocity with typical reverse flow and very high impedance of flow (Figures 3.6–3.8). *The common iliac vessels* can only occasionally be seen because they are usually too far from the probe (Figure 3.9). If seen, they present a high velocity of blood flow with the most prominent reverse flow in diastole. The phenomenon of reversed flow in the iliac vessels, familiar to physicists, is much less well known to physicians. The high peripheral resistance in the pelvis and the legs causes blood to rebound back up to the iliac arteries and the aorta.

Table 3.1 Success rate of visualization of pelvic vessels by transvaginal color Doppler ($n = 300$)

	Women		
Vessel	*Healthy* (*n = 150*)	*Postmenopausal* (*n = 50*)	*Pregnant* (*n = 100*)
External iliac artery	142 (94.7%)	39 (78.0%)	89 (89.0%)
Internal iliac artery	149 (99.3%)	41 (82.0%)	93 (93.0%)
Uterine artery	150 (100%)	40 (80.0%)	100 (100%)
Ovarian artery	32 (21.3%)	2 (4.0%)	14 (14.0%)

Of great interest to gynecologists are the visualization and analysis of the uterine and ovarian vessels. *The uterine artery* originates from the internal iliac artery, whereas the ovarian artery is a direct branch of the abdominal aorta. The uterus has an extensive vascular network. The color Doppler signal from the main uterine vessels may be seen in all patients laterally to the cervix at the level of the cervicocorporeal junction of the uterus (Figures 3.10–3.13). The small branches of the uterine artery can be followed by searching corpus uteri ascending along the lateral wall. Even small, terminal branches can be visualized in some patients in the direction towards the ovary or to the myometrium (Figure 3.14). Waveform analysis shows high to moderate velocity (Figures 3.15–3.17). The resistance index depends on age, the phase of the menstrual cycle and any special conditions (e.g. pregnancy, tumor) and is usually very high (Table 3.2)[3–5]. There is no significant difference in the resistance index to blood flow between the left and right uterine arteries, which would indicate that blood is equally distributed to both sides of the uterus[6]. Other evidence also supports this. Injection studies performed on specimens resected at operation or at autopsy have shown that dye penetrates uniformly through the endometrium[3].

The increase in blood flow to the uterus in response to estrogens is well documented[7, 8]. In general, it is maximum when estrogen levels are high, which is

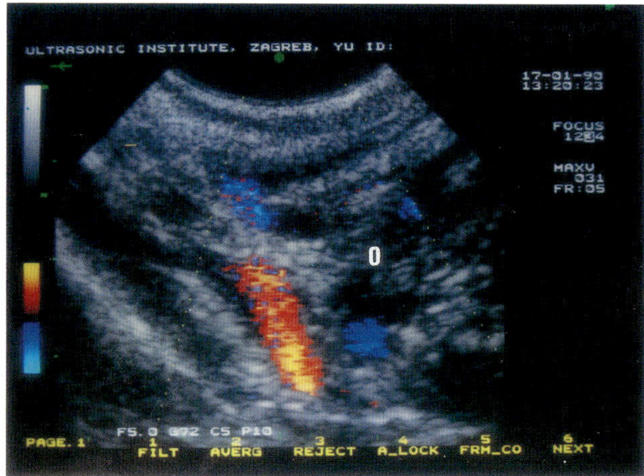

Figure 3.1 The external iliac vessel visualized by transvaginal color Doppler. o = right ovary

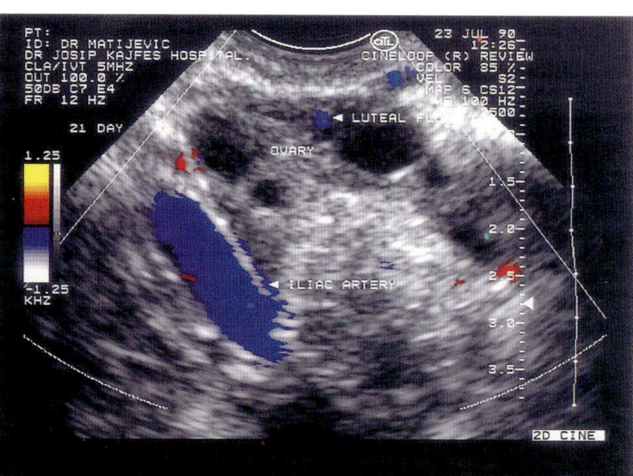

Figure 3.2 Another example of blood flow easily detected in external iliac vessels by color Doppler imaging

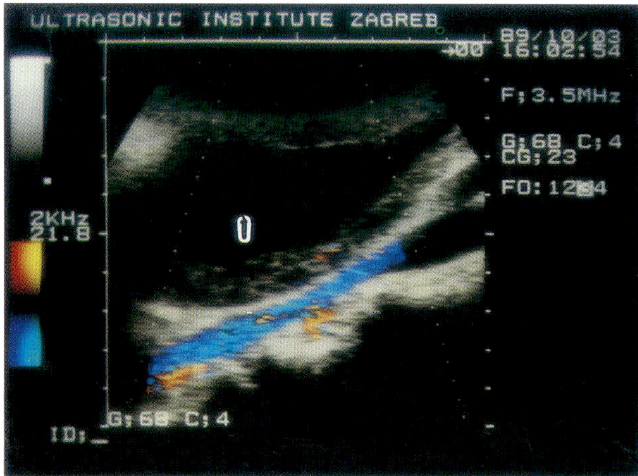

Figure 3.3 An oblique scan of the ovary. Color flow represents internal iliac vessels. o = ovary

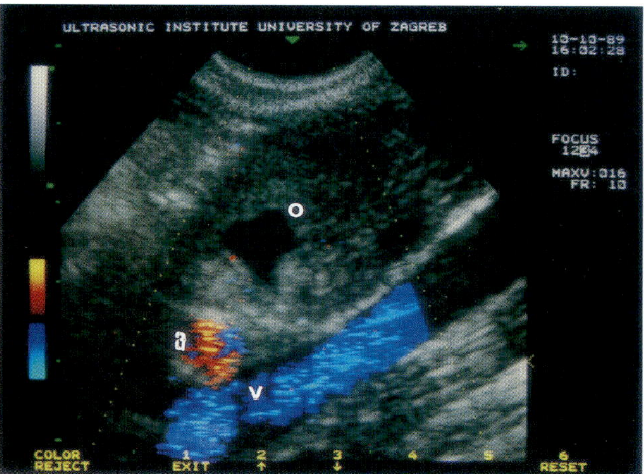

Figure 3.4 Color Doppler representation of the internal iliac vessels. The brightness of the color corresponds to the velocity of blood flow. a = internal iliac artery; v = internal iliac vein; o = ovary

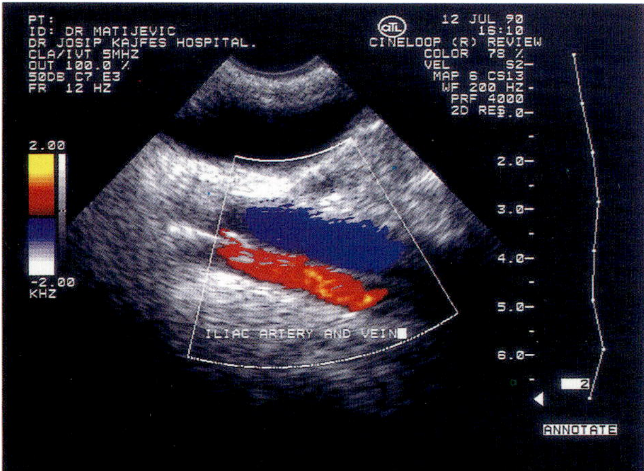

Figure 3.5 Another example of internal iliac blood flow detected by color Doppler

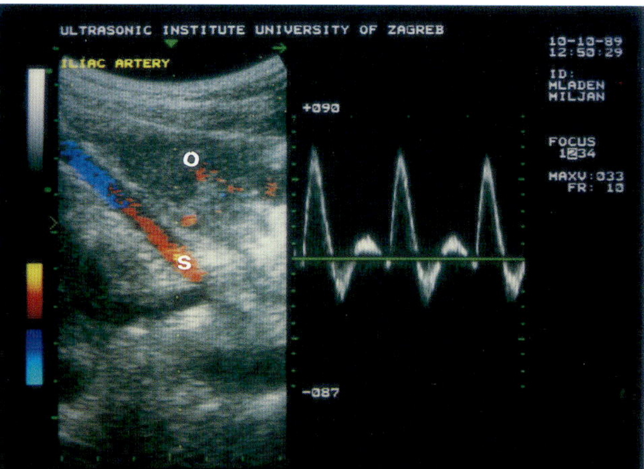

Figure 3.6 The external iliac artery. Pulsed Doppler waveform (right) shows high-velocity and high-resistance flow with characteristic normal finding of reverse flow. s = sample volume; o = right ovary

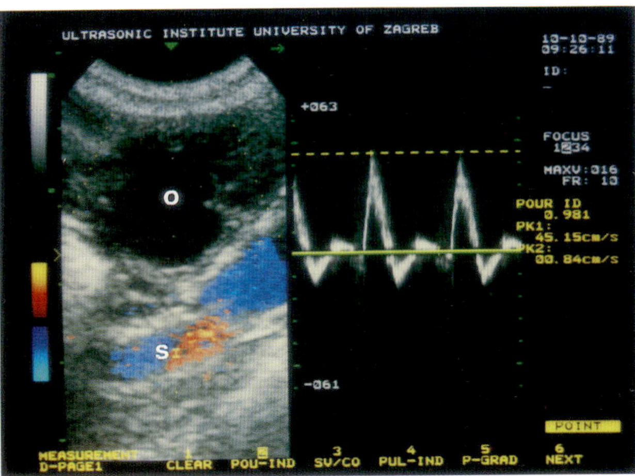

Figure 3.7 The internal iliac artery. A similar Doppler pattern to that in the case of the external iliac artery. These are the normal findings of blood flow through the large pelvic artery. s = sample volume; o = ovary

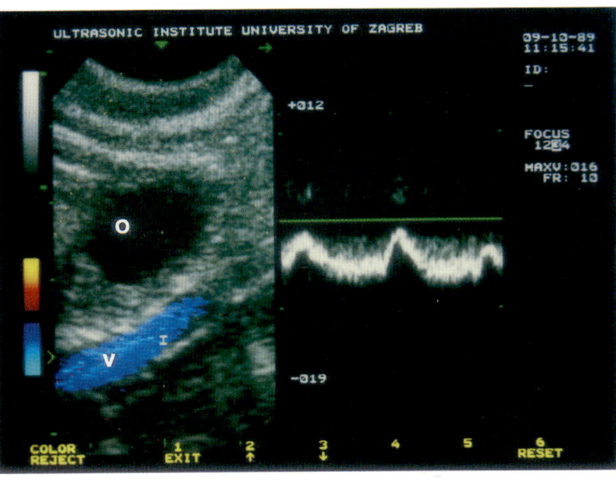

Figure 3.8 The internal iliac vein. This shows a very low-velocity blood flow without the regular systolic–diastolic variation. The continuous venal flow pattern is affected by the transmitted pulsations of the internal iliac artery. v = internal iliac vein; o = ovary

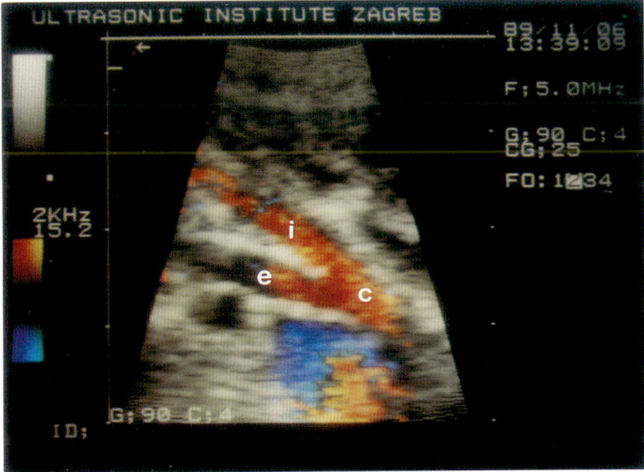

Figure 3.9 Montage showing the bifurcation of the common iliac artery into the internal and external iliac arteries. c = common iliac artery; e = external iliac artery; i = internal iliac artery

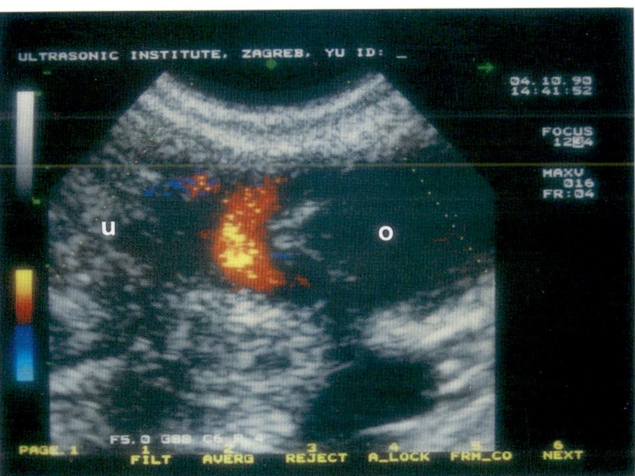

Figure 3.10 The color Doppler signal from the main uterine artery seen laterally to the cervix at the level of the cervicocorporeal junction of the uterus. u = uterus; o = ovary

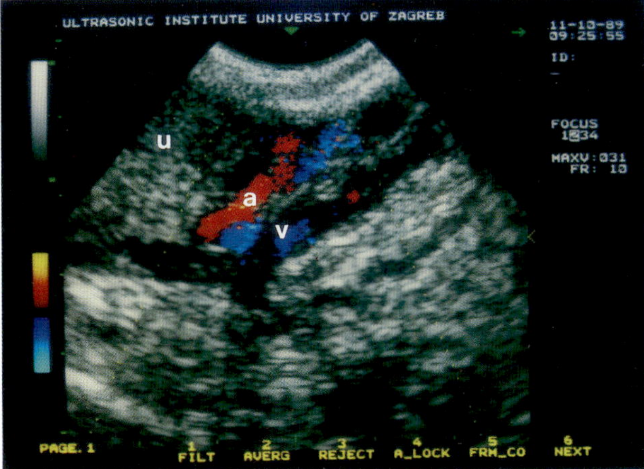

Figure 3.11 Another example of color-coded blood flow of both uterine vessels. a = uterine artery; v = uterine vein; u = uterus

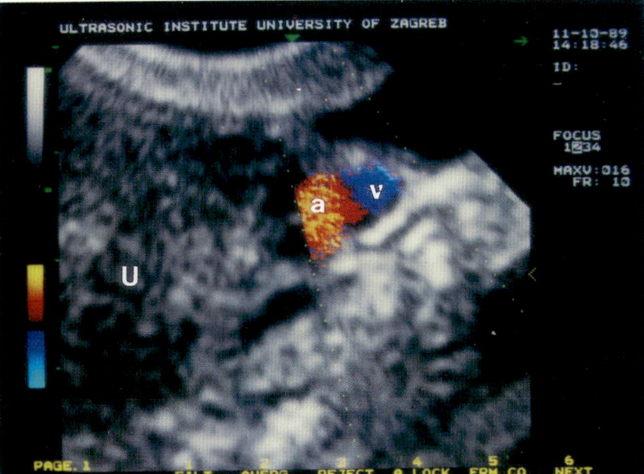

Figure 3.12 A cross-section through the uterine artery and vein. a = uterine artery; v = uterine vein; u = uterus

Table 3.2 Mean resistance index (± SD) of blood flow in the uterine artery ($n = 290$)

		Resistance index	
Patients	*n*	*Left uterine artery*	*Right uterine artery*
Healthy fertile	150	0.86 ± 0.04	0.85 ± 0.07
Postmenopausal	40	0.89 ± 0.06	0.90 ± 0.05
Pregnant	100	0.76 ± 0.07	0.75 ± 0.05
Total	290	0.84 ± 0.07	0.83 ± 0.06

Table 3.3 Mean resistance index (± SD) of blood flow in the uterine artery of healthy fertile women in relation to the menstrual cycle ($n = 150$)

Cycle	*n*	*Resistance index*
Proliferative phase	130	0.88 ± 0.05
Luteal phase	120	0.84 ± 0.06

$t = 5.70$; $p < 0.01$

during the week preceding spontaneous ovulation and during the secretory phase of the menstrual cycle. The resistance index in the proliferative phase is, in normal circumstances, significantly higher than in the luteal phase (Table 3.3). Such findings could play an important role in clinical assessment of uterine perfusion in infertile women (see Chapter 4). The effects of pregnancy on the uterine circulation are also well documented[5, 9–16]. Doppler shift waveforms indicate a low impedance of blood flow.

The ovarian arteries arise from the upper aorta, pass down the retroperitoneum, and enter the pelvis through the infundibulopelvic ligament on the side wall of the pelvis. The ovarian vessels can often be seen in this position (Figure 3.18). The angle of the transducer is manually adjusted to optimize the amplitude of the Doppler signal, which is monitored both by ear and by the waveform on the spectrum analyzer. Color flow is usually not prominent, velocity is low and resistance is very variable according to the phase of the menstrual cycle. This functional activity affects volume flow to the ovary and increases the amplitude of the signal, since the amplitude varies with the number of reflectors (red cells). Thus, an inactive ovary is a difficult organ from which to elicit signals, whereas an active ovary gives rise to signals that are easily detected[17]. Basically, the functional activity of the ovaries is of particular importance in infertility and the clinical assessment of its therapy. The possibility that ovarian activity may be associated with lowered vascular impedance and hence modify the color Doppler signal merits further study.

UTERINE AND OVARIAN PERFUSION

Vascularization of normal ovarian and uterine tissue could not be visualized until transvaginal color Doppler came into use. In normal conditions, uterine and ovarian perfusion is moderate in healthy fertile women; it is highly reduced and not prominent in healthy postmenopausal women, as a normal consequence of the involutive process. The rich perfusion is normally obvious in pregnant women. Such small and randomly dispersed vessels present low velocity and moderate impedance blood flow (Figures 3.19 and 3.20). The ovarian blood flow depends on the phase of the menstrual cycle. In the proliferative part of the menstrual cycle, the flow usually is not prominent. Contrary to this, the luteal phase is shown by detectable color due to the *corpus luteum* (Figures 3.21–3.24). Angiogenesis in the corpus luteum occurs in physiological circumstances each menstrual cycle. These newly formed vessels may show prominent blood flow detected by transvaginal color Doppler. Noted changes are derived from a tremendous decrease of the vascular impedance and, therefore, an increase of the diastolic blood flow velocity[18]. Such low-impedance flow should not be mistaken for tumor neovascularization or ectopic pregnancy, which is characterized by low impedance but moderate or high velocity of blood flow[19].

Taylor showed that, early in the cycle (days 1–7), both ovaries gave similar waveforms, consisting of a high-impedance signal of low amplitude and without evidence of diastolic flow[17]. By day 9, the ovary developing the dominant follicle showed diastolic flow and also increased amplitude, making it easier to detect. The amplitude in that ovary continued to increase through ovulation and into the functioning period of the corpus luteum, becoming maximal around day 21. After day 23, the waveform rapidly changed back to the inactive, high-impedance pattern seen on day 1. Through this cycle, the contralateral ovary showed the same low-amplitude, high-impedance signals without diastolic flow. Usually, the inactive ovary becomes active the following month.

Prior to the development of Doppler technology, there had been a few studies of ovarian flow in humans. However, there have been many previous reports on luteal flow in domestic and farm animals. Blood flow has been shown to vary during the estrous cycle in the ewe, cow and other mammals, being highest during the luteal phase and lowest during the follicular phase[20, 21]. These investigations have shown that 80–90% of the blood flow to the ovary is supplied to the corpus luteum. In rabbits and sheep, luteal blood flow has been shown to be more than 10 times that of surrounding tissues[22, 23]. This increased blood flow results from the production of new blood

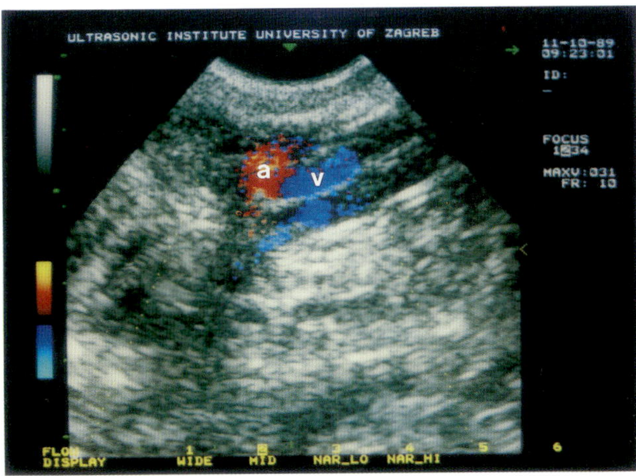

Figure 3.13 An oblique scan of the uterine vessels. a = uterine artery; v = uterine vein

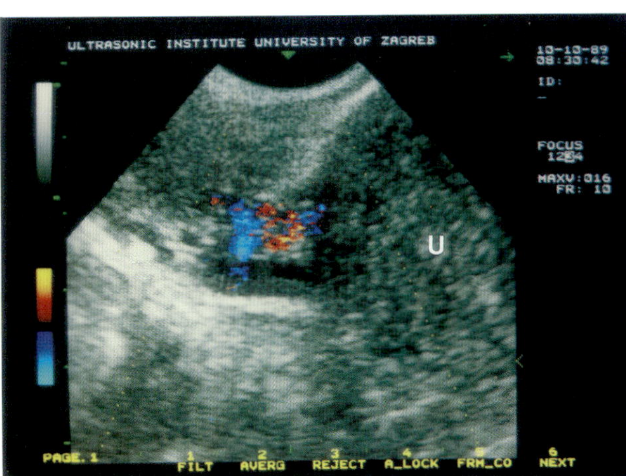

Figure 3.14 Color flow imaging of the small, terminal branches of the uterine artery, in the direction towards the myometrium. u = uterus

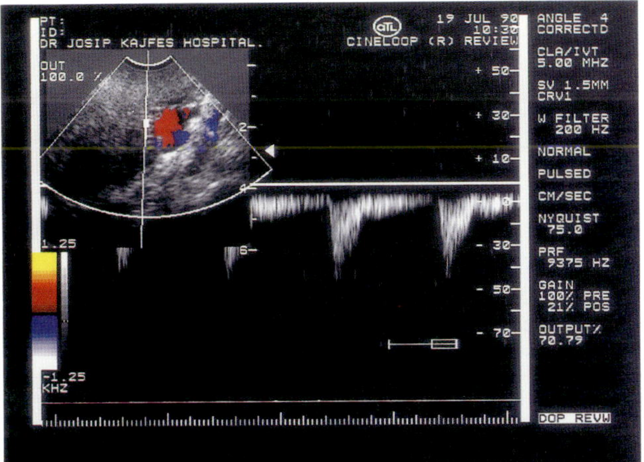

Figure 3.15 The uterine artery. Pulsed Doppler (right) showing high-to-moderate velocity and a high resistance of blood flow. This is the normal finding for uterine artery blood flow

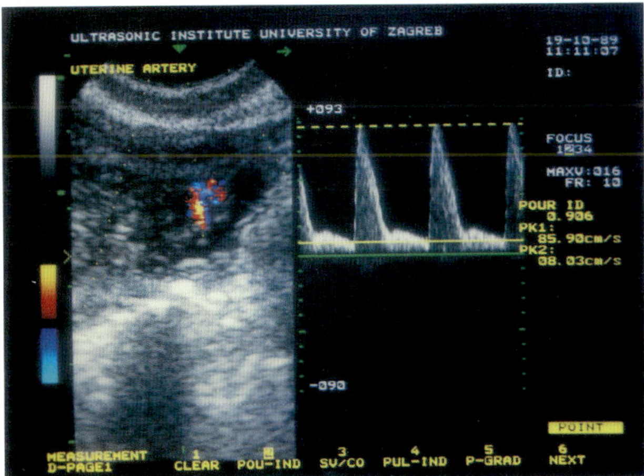

Figure 3.16 High-velocity and high-resistance (RI = 0.906) blood flow of the uterine artery. These are characteristic findings in the proliferative part of the menstrual cycle

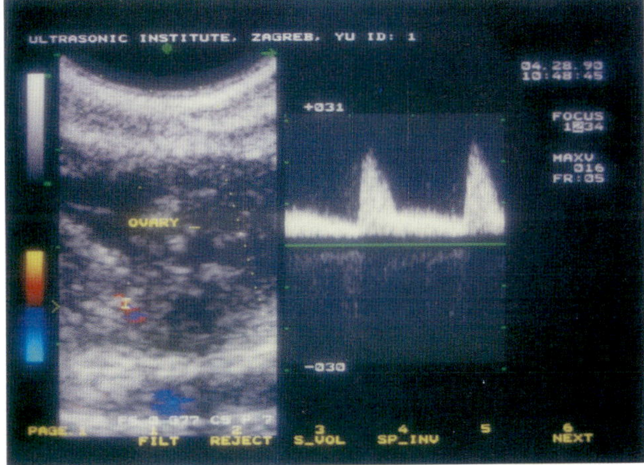

Figure 3.17 The uterine artery. An increased diastolic flow, indicating the luteal part of the cycle. A similar pattern can be obtained in the case of pregnancy

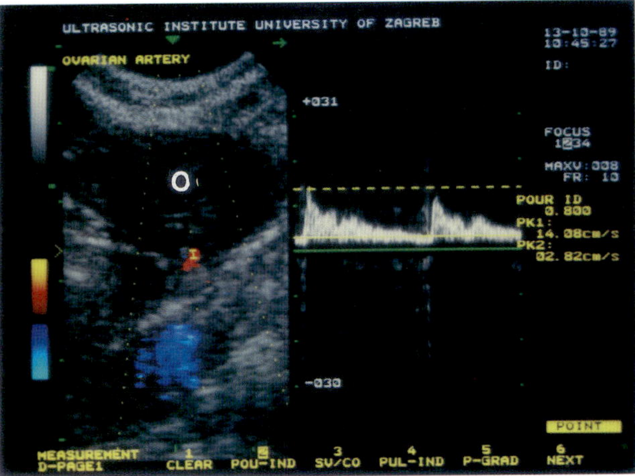

Figure 3.18 Ovarian artery blood flow. A typical low-velocity pulsed Doppler signal (right) is obtained. The presence of a diastolic component of flow indicates the luteal phase of the cycle. o = ovary

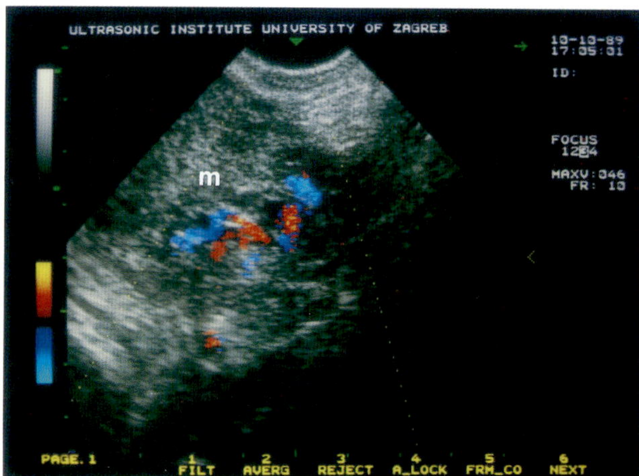

Figure 3.19 Color Doppler showing small and randomly dispersed myometrial vessels in the case of pregnancy. m = myometrium

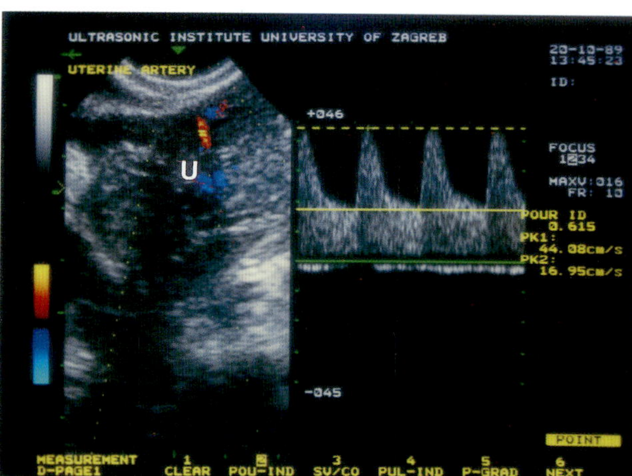

Figure 3.20 An increased diastolic flow of the uterine artery indicating rich vascularization of the uterus. Normal finding in the case of pregnancy. u = uterus

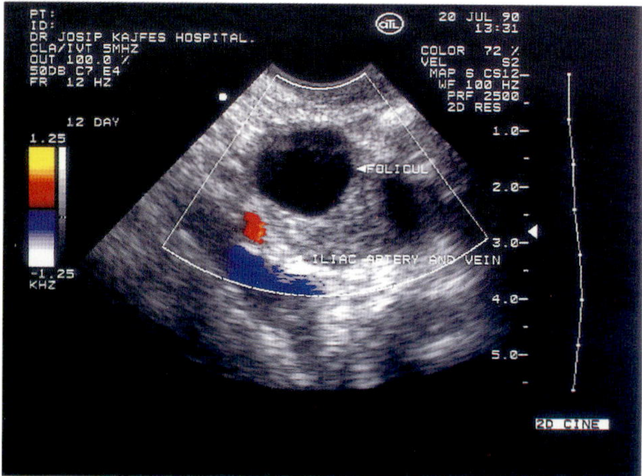

Figure 3.21 Developing follicles in the case of an infertile patient. Color flow on the periphery of the preovulatory follicles indicates their maturation

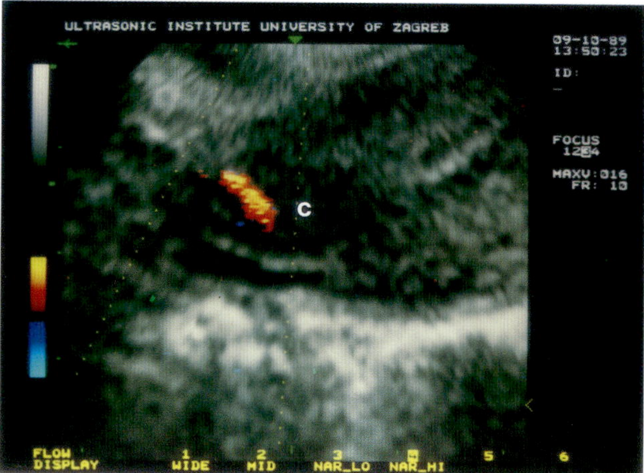

Figure 3.22 Neovascularization of the corpus luteum detected by color Doppler. c = corpus luteum

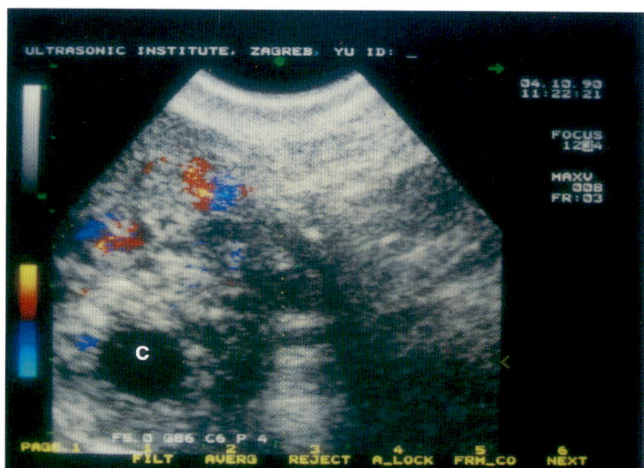

Figure 3.23 A color-coded blood supply of the corpus luteum cyst. c = corpus luteum cyst

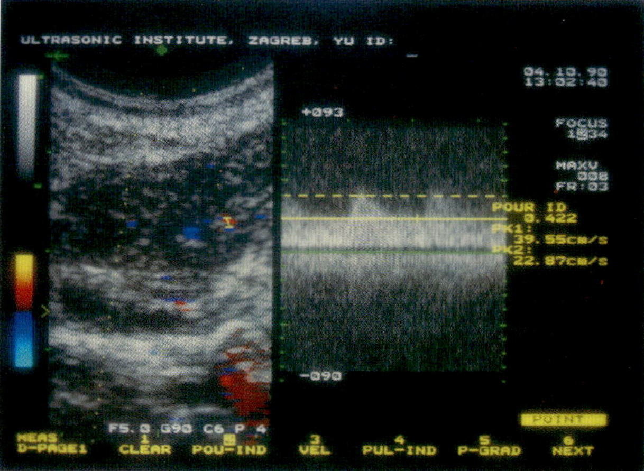

Figure 3.24 The same patient. Pulsed Doppler analysis (right) shows moderate velocity and increased diastolic flow. There is a typical very low resistance (RI = 0.422) blood flow characteristic of newly formed vessels of the corpus luteum

vessels, similar to those observed around the growing edges of tumors. This high blood flow is necessary for the supply of the precursors of steroidogenesis and for the removal of synthesized progesterone. There is a rapid decrease in ovarian blood flow accompanying luteolysis[22, 24]. Prostaglandin $F_{2\alpha}$ is reported to be the active luteolysin, and a reduction of 90% in luteal flow has been reported following administration of this prostaglandin[25].

In a longitudinal study through the physiological menstrual cycle, decreased vascular impedance, implying increased flow, occurred at an early stage of the human menstrual cycle, before the dominant follicle could be recognized by other methods[18]. Ovarian steroids alter the function of periarterial sympathetic nerves through changes in α_1-adrenergic receptor numbers, and lead to marked changes in the blood flow[26]. In animal studies, the injection of estradiol led to an increase of blood flow in pelvic arteries[27]. In normal cycles, the end-diastolic flow in the ovary with the developing follicle is over five times the threshold of the lowest detectable frequency[2]. In patients with stimulated cycles and bilateral ovulating ovaries, these active changes with low-impedance signals are seen bilaterally. If pregnancy occurs, then the corpus luteum maintains its function for the first trimester of pregnancy. Pregnancy is known to be dependent on a functioning corpus luteum and Doppler ultrasound is an excellent means by which to identify this activity. A most important conclusion follows: the absence of a functioning corpus luteum could indicate a non-viable pregnancy, whether normotopic or ectopic[17]. Therefore, the ability to display luteal flow is important.

CONCLUSION

Transvaginal color Doppler study demonstrates the great potential for investigation of pelvic vascular physiology. The applications of this technique are rapidly widening and may soon be applied to the monitoring of patients being treated for tumors or infertility. However, the important field of assessing the normal hemodynamic state of the uterus and ovaries, as an aid to understanding pelvic pathology, has been rather neglected. It is hoped that the results presented in this chapter will act as a comparison to other studies using transvaginal color Doppler sonography in the pelvis to delineate normal and abnormal pelvic blood flow.

REFERENCES

1. Taylor, K.J.W., Burns, P.N., Woodcock, J.P. and Wells, P.N.T. (1985). Blood flow in deep abdominal and pelvic vessels: ultrasonic pulsed-Doppler analysis. *Radiology*, **154**, 487
2. Taylor, K.J.W., Burns, P.N., Wells, P.N.T., Conway, D.L. and Hull, M.G.R. (1985). Ultrasound Doppler flow studies of the ovarian and uterine arteries. *Br. J. Obstet. Gynaecol.*, **92**, 240
3. Long, M.G., Boultbee, J.E., Hanson, M.E. and Beget, R.H.J. (1989). Doppler time velocity waveform studies of the uterine artery and uterus. *Br. J. Obstet. Gynaecol.*, **96**, 588
4. Goswamy, R.K. and Steptoe, P.C. (1988). Doppler ultrasound studies of the uterine artery in spontaneous ovarian cycles. *Hum. Reprod.*, **3**, 721
5. Schulman, H., Fleischer, A., Farmakides, G., Bracero, L., Rochelson, B. and Grunfeld, L. (1986). Development of uterine artery compliance in pregnancy detected by Doppler ultrasound. *Am. J. Obstet. Gynaecol.*, **155**, 1031
6. Kurjak, A., Jurković, D., Alfirevic, Z. and Žalud, I. (1990). Transvaginal color Doppler imaging. *J. Clin. Ultrasound*, **18**, 227
7. Williams, M.F. (1948). The vascular architecture of the rat uterus as influenced by estrogen and progesterone. *Am. J. Anat.*, **83**, 247
8. Greiss, F.C. and Anderson, S.G. (1970). Effect of ovarian hormone on the uterine vascular bed. *Am. J. Obstet. Gynecol.*, **107**, 829
9. Campbell, S., Griffin, D.R., Pearce, J.M. *et al.* (1983). New Doppler technique for assessing uteroplacental blood flow. *Lancet*, **1**, 675
10. Deutinger, J., Rudelstorfer, R. and Bernaschek, G. (1988). Vaginosonographic velocimetry of both main uterine arteries by visual vessel recognition and pulsed Doppler method during pregnancy. *Am. J. Obstet. Gynecol.*, **159**, 1072
11. Cohen-Overbeek, T., Pearce, J.M. and Campbell, S. (1985). The antenatal assessment of utero-placental and feto-placental blood flow using Doppler ultrasound. *Ultrasound Med. Biol.*, **11**, 329
12. Trudinger, B.J., Giles, W.B. and Cook, C.M. (1985). Uteroplacental blood flow velocity waveforms in normal and complicated pregnancy. *Br. J. Obstet. Gynaecol.*, **92**, 39
13. Kurjak, A., Alfirevic, Z. and Miljan, M. (1988). Conventional and color Doppler in the assessment of fetal and maternal circulation. *Ultrasound Med. Biol.*, **14**, 337
14. Kurjak, A., Žalud, I. and Crvenković, G. (1989). Transvaginal color Doppler in the assessment of maternal and fetal circulation. Proceedings of International Symposium: *Transvaginal Sonography*, Rotterdam, November 2–4, p. 93 (abstr.)
15. Kurjak, A., Miljan, M., Jurković, D., Alfirevic, Z. and Žalud, I. (1989). Color Doppler in the assessment of fetomaternal circulation. *Rech. Gynecol.*, **1**, 269
16. Alfirevic, Z. and Kurjak, A. (1990). Transvaginal color and pulsed wave Doppler in the assessment of blood flow in the first trimester of pregnancy. *J. Perinat. Med.*, **18**, 173
17. Taylor, K.J.W. (1988). Pulsed Doppler ultrasound of the pelvis and the first trimester of pregnancy. In Taylor, K.J.W., Burns, P.N. and Wells, P.N.T. (eds.) *Clinical Applications of Doppler Ultrasound*, p. 246. (New York: Raven Press)

18. Deutinger, J., Reinthaller, A. and Bernaschek, G. (1989). Transvaginal pulsed Doppler measures of blood flow velocity in the ovarian arteries during cycle stimulation and after follicle puncture. *Fertil. Steril.*, **51**, 466
19. Žalud, I. and Kurjak, A. (1990). The assessment of luteal blood flow in pregnant and non-pregnant women by transvaginal color Doppler. *J. Perinat. Med.*, **18**, 215
20. Ford, S.P. and Chenault, J.R. (1981). Blood flow to the corpus luteum-bearing ovary and ipsilateral uterine horn of cows during oestrous cycle and early pregnancy. *J. Reprod. Fertil.*, **62**, 555
21. Hossain, M.I., Lee, C.S., Clarke, I.J. and O'Shea, J.D. (1979). Ovarian and luteal blood flow and peripheral plasma progesterone levels in cyclic guinea-pigs. *J. Reprod. Fertil.*, **57**, 167
22. Bruce, N.W. and Moor, T.J. (1976). Capillary blood flow to ovarian follicles, stroma and corpora lutea of anesthetized sheep. *J. Reprod. Fertil.*, **46**, 299
23. Novy, M.J. and Cook, M.J. (1973). Redistribution of blood flow by prostaglandin $F_{2\alpha}$ in the rabbit ovary. *Am. J. Obstet. Gynecol.*, **117**, 381
24. Magness, R.R., Christenson, R.K. and Ford, S.P. (1983). Ovarian blood flow throughout the estrous cycle and early pregnancy in sows. *Biol. Reprod.*, **28**, 1090
25. Nett, T.M. and Niswender, G.D. (1981). Luteal blood flow and receptors for LH during $PGF_{2\alpha}$-induced luteolysis: production of $PGF_{2\alpha}$ during early pregnancy. *Acta Vet. Scand.* (Suppl.), **77**, 117
26. Ford, S.P., Reynolds, L.P. and Farley, D.B. (1984). Interaction of ovarian steroids and periarterial alpha-1-adrenergic receptors in altering uterine blood flow during the estrous cycle of gilts. *Am. J. Obstet. Gynecol.*, **150**, 480
27. Randall, N.J., Beard, R.W., Sutherland, I.A., Fugueroa, J.P., Drost, C.J. and Nathanielsz, P.W. (1988). Validation of thermal techniques for measurement of pelvic organ blood flows in the nonpregnant sheep: comparison with transit-time ultrasonic and microsphere measurements of blood flow. *Am. J. Obstet. Gynecol.*, **158**, 651

4 Infertility

A. Kurjak and S. Kupešić-Urek

Today ultrasound is one of the most important techniques for the study of reproductive disorders because of its non-invasiveness, safety and low cost[1]. Until recently, transabdominal sonography was the standard method used for studying follicular growth and maturation. For clear visualization of the pelvic anatomy by this method a full bladder is required. The great practical advantage of vaginal sonography is that it obviates the need for a full bladder. It also provides better resolution because the proximity of the probe to the pelvic structures permits the use of higher ultrasound frequencies.

Blood flow studies of the pelvic circulation represent the most recent development in ultrasound assessment of female infertility. The pelvic circulation can be easily analyzed by combining real-time imaging with pulsed Doppler[2,3].

FOLLICULAR GROWTH, OVULATION AND CORPUS LUTEUM

Ultrasound findings

Monitoring of follicular growth and ovulation by ultrasound is usually attempted from day 9 to day 10 of a regular 28-day menstrual cycle. The dominant follicle can be identified from day 9 to day 10 of a regular 28-day cycle as an anechoic cystic structure with sharp borders, usually measuring 8–10 mm in diameter. The dominant follicle continues to grow at a speed of 2–3 mm per day[4–7] (Figure 4.1). This increase in follicular size parallels the rising estradiol levels which occur in the menstrual cycle at this time[8,9].

Changes within the preovulatory follicle (Figure 4.2) include luteinization of granulosa cells, expansion and dissociation of the cumulus oophorus, separation of granulosa and theca layers and, finally, oocyte maturation. Dissociation of the cumulus oophorus occurs 20 hours before ovulation[9].

The visualization of the cumulus oophorus as a triangular intrafollicular echogenic structure is a very reliable sign of forthcoming ovulation[10]. Rupture of the preovulatory follicle and ovulation are results of biochemical changes. It seems that increased prostaglandin synthesis plays an important role in the mechanism of follicular rupture[11]. During ovulation, antral fluid from the follicle is released and echogenic areas appear within the follicle[7]. Complete collapse of the follicle takes place within 7–35 minutes. As early as 1 hour after ovulation, the corpus hemorrhagicum may be seen[12,13]. The diagnosis of ovulation can be made reliably in cases where there is a completely collapsed follicle or a residual thick-wall cyst usually filled with echogenic blood clots[14]. A subjective impression of an increase in the amount of fluid in the pouch of Douglas may also be used to confirm the diagnosis of ovulation[15].

Transvaginal color Doppler findings

On transvaginal color Doppler, ovarian follicles can be easily differentiated from other anatomical structures such as cross-sections of the internal iliac vessels or other nearby blood vessels. Hydrosalpinx and an ovarian cyst can be differentiated from ovarian follicles by serial daily transvaginal sonography scans since they present a constant appearance.

Follicular flow can be detected when the dominant follicle reaches 12–15 mm in diameter and it may be an important hemodynamic parameter of its growth, maturation and ovulation (Figures 4.3–4.7).

After ovulation, conventional Doppler signals can be correlated with the development of corpus hemorrhagicum and the growth and progression of corpus luteum (Figures 4.8–4.10). The corpus luteum is usually seen a few days after ovulation and, because of its echodensity, it is easily distinguishable from the surrounding stroma[2,7].

Color flow from the ovarian tissue in infertility patients, undergoing serial transvaginal color Doppler examination for follicular, endometrial and pelvic circulation dynamics, is usually obtainable from the 10th day of cycle until ovulation, as well as in the luteal phase. In most patients early luteal color flow from ovarian tissue was explored and the resistance index (RI) was calculated (RI = 0.44 ± 0.06). In the middle of the luteal phase the corpus luteum was mainly seen as an echogenic area or as a cyst.

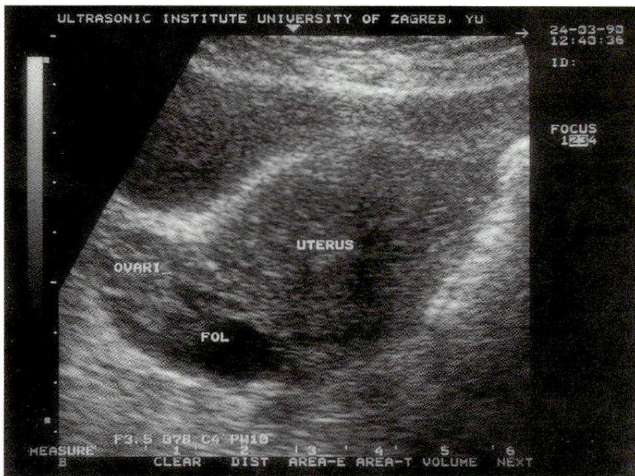

Figure 4.1 The transverse scan performed on the 12th day of the cycle. The right ovary contains the dominant follicle

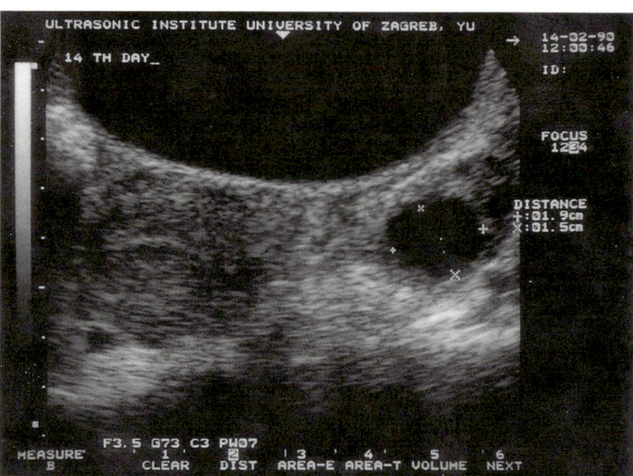

Figure 4.2 The preovulatory follicle is visible in the left ovary on the 14th day of the cycle. The endometrium exhibits increased echogenicity

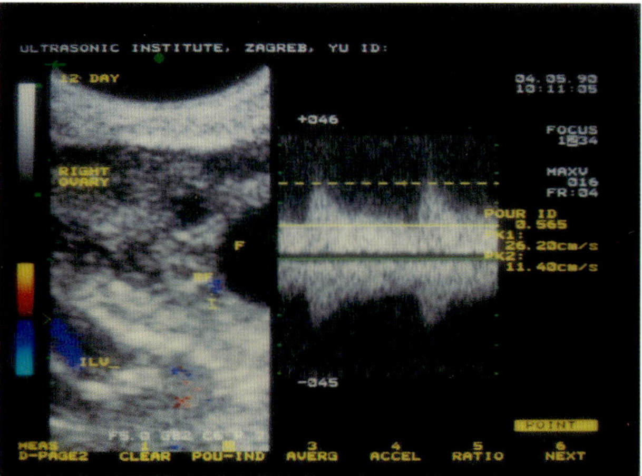

Figure 4.3 Demonstration of the follicular blood flow on the 12th day of a 30-day regular menstrual cycle (left). Flow velocity waveforms were obtained around the follicle (right)

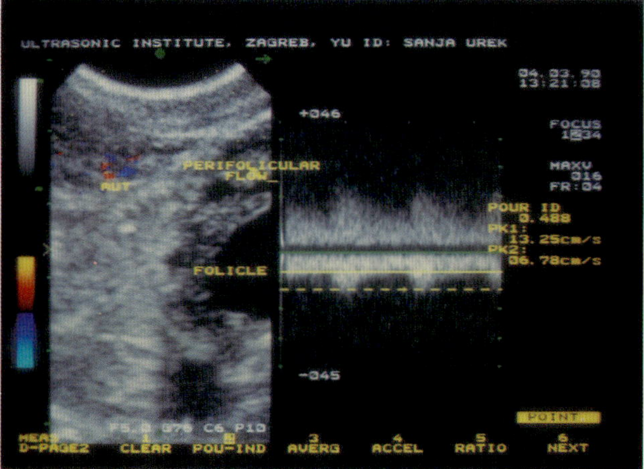

Figure 4.4 On the following day the size of the dominant follicle increased (left) and flow velocity waveforms were obtained (right)

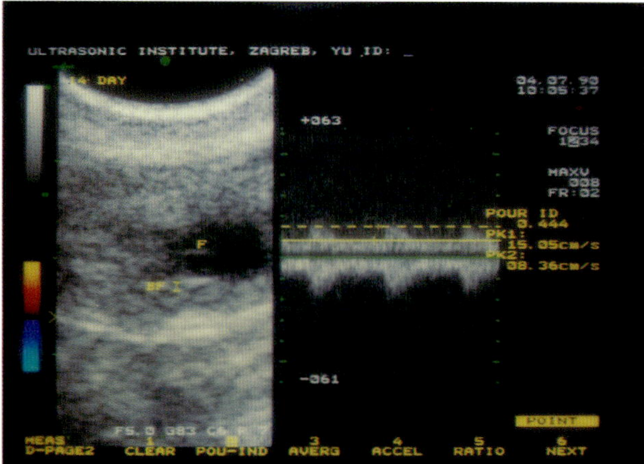

Figure 4.5 The same patient on the 14th day. The mean diameter of the dominant follicle reached 15 mm

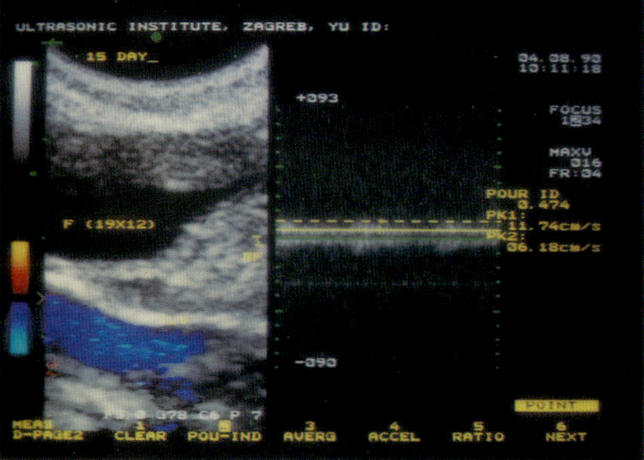

Figure 4.6 On the following day the size of the dominant follicle measured 19 x 12 mm

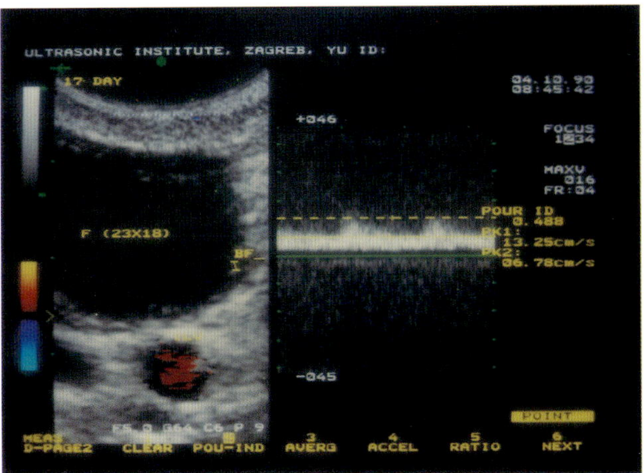

Figure 4.7 The preovulatory follicle on the 17th day of the cycle (right). A low-impedance flow velocity waveform (right) was obtained from the follicle

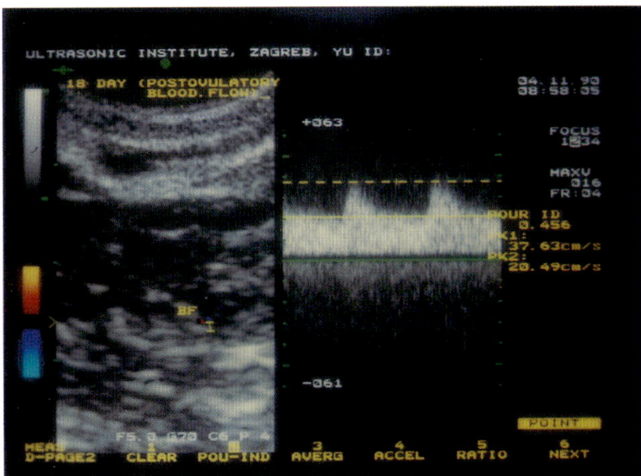

Figure 4.8 The sonogram of the same patient on the following day shows a fresh corpus luteum (left). The waveform analysis is on the right

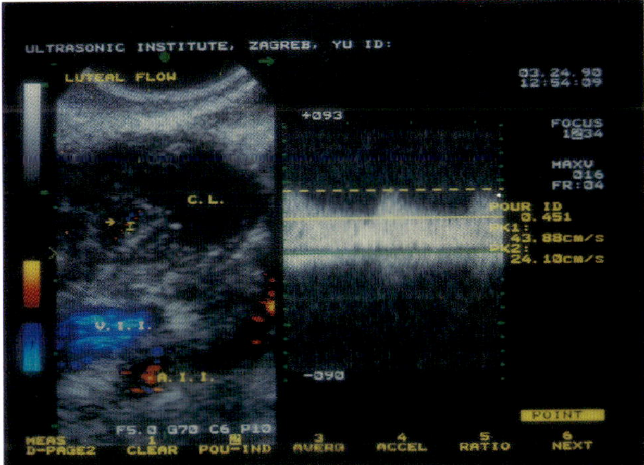

Figure 4.9 The sonogram of the same patient 2 days later

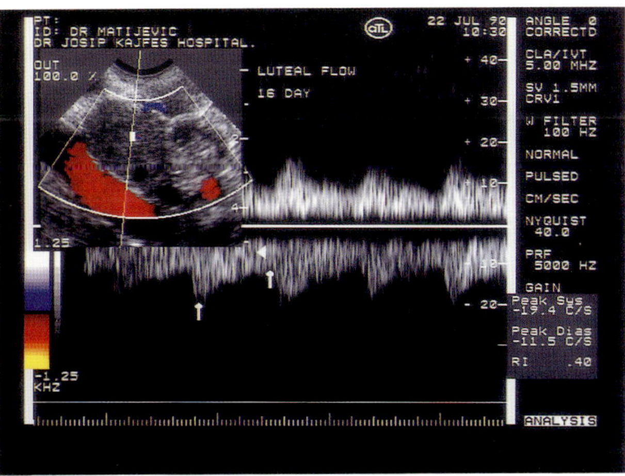

Figure 4.10 Demonstration of increased vascularity in the corpus luteum as seen by color Doppler 2 days after ovulation (left). A low-impedance flow velocity waveform (right) was obtained from the corpus luteum

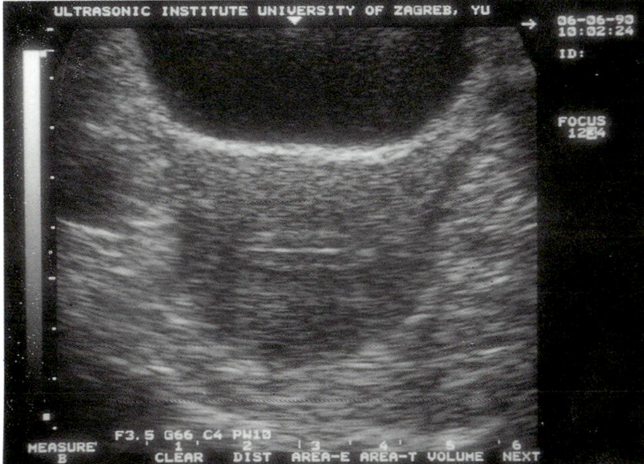

Figure 4.11 The transversal scan showing the endometrium in the late proliferative phase. The endometrium is hypoechoic compared to the myometrium. The central echo is also clearly visible

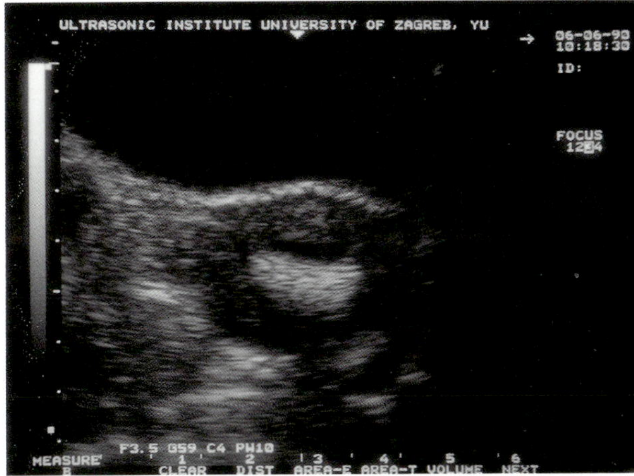

Figure 4.12 Echogenic endometrium in the secretory phase. The endometrium is thick and the central echo is lost

Corpus luteum flow was detected in all ovulatory cycles. The resistance index of corpus luteum flow in cases of early intrauterine pregnancy was 0.53 ± 0.09. In patients with proven ectopic pregnancy, the resistance index of luteal flow was 0.48 ± 0.07. The resistance index of luteal flow was found in the case of non-pregnant women to be 0.42 ± 0.12. Since an increasing resistance index describes an inadequacy of the corpus luteum which is incompatible with pregnancy, it might become an accurate predictor for the evaluation of a normal or pathological pregnancy. Investigations in the luteal phase might show a new aspect for the diagnosis and treatment of luteal phase insufficiency[16, 17].

The monitoring of ovulation is important because early complications such as hyperstimulation can be easily detected by this method.

ENDOMETRIAL CHANGES

Ultrasound findings

The endometrium, representing the end-organ for ovarian-secreted estrogen and progesterone, changes in appearance during the normal ovulatory cycle[18]. In the immediate postmenstrual phase, the endometrium is seen as a bright thin line of about 5 mm in thickness. During the second week of the proliferative phase it becomes thicker, with a characteristically thin and well-defined boundary in the basal layer (Figure 4.11). In the preovulatory phase of the cycle the endometrium, measured by ultrasound, is 3.5–7 mm thick. The central echo is now less pronounced, while a highly echogenic linear echo is present in its basal portions[19]. This characteristic ultrasonic appearance of the preovulatory endometrium is well known as the 'ovulation ring' and it can be used as an additional sign of ovulation in spontaneous cycles[20]. In the mid-luteal phase, the endometrium becomes more even and represents secretory endometrium[7] (Figure 4.12). The largest increase in endometrial thickness occurs between the early and the late proliferative stages, and not between the proliferative and secretory phases[21].

Adams and colleagues have shown a close correlation between estradiol levels in the late follicular phase of the cycle and endometrial thickness[22]. No detectable changes have been reported during the luteal phase which would allow sonographic evaluation of the endometrium to predict whether pregnancy will or will not occur[23-26].

Transvaginal color Doppler findings

The technique of transvaginal color Doppler offers the possibility of studying the alterations of endometrial blood flow under physiological and pathophysiological conditions (Figure 4.13).

ANOVULATORY CYCLES

Ultrasound remains particularly helpful in making the diagnosis of polycystic ovary syndrome. The sonographic image of polycystic ovaries is typical and seems pathognomonic. A large number of small cystic structures are 'crowded' and stand out from the surface of enlarged ovarian stroma[27, 28] (Figure 4.14).

Daily ultrasound examination facilitates the detection of abnormalities of follicular development and ovulation, such as the luteinized unruptured follicle syndrome[29, 30] (Figure 4.15). The main characteristic of this condition is complete luteinization of the follicle without ovulation and oocyte release, but the etiology still remains unclear. Ultrasound monitoring of follicular growth performed on a daily basis shows luteinization of the unruptured follicle, seen as progressive accumulation of strong echoes primarily located on its periphery[31, 32].

BLOOD FLOW STUDIES OF OVARIAN AND UTERINE CIRCULATION

As described earlier in this book, pelvic circulation can be analyzed by using real-time imaging and pulsed Doppler[33]. Characteristic flow velocity waveforms of external and internal iliac, uterine and ovarian arteries were first described by using transabdominal Doppler[2, 3].

The pulsed and color Doppler transvaginal probe will become the technique of choice for scanning the pelvic circulation in female infertility because of the small distance between the probe and the vessels, giving better resolution, and because of the improved convenience for the patient, since a full bladder is not necessary for this examination.

The use of blood flow studies in infertility opens a new and interesting avenue in the investigation of physiological changes as well as pelvic pathology. Transvaginal color Doppler simplifies the examination technique, facilitates the examination of small vascular branches, and increases the reproducibility of measurements.

A color Doppler signal from both uterine arteries can be easily seen lateral to the cervix[34]. The waveform classification of these arteries is based on the absence or presence and duration of diastolic flow. It is noted that diastolic and/or end-diastolic flow are not present in every infertile patient[35, 36] (Figure 4.16). Goswamy and colleagues have suggested that poor uterine blood flow is a cause of infertility. The aims of that study[35] were to assess uterine perfusion response in infertile women during spontaneous ovarian cycles, and to assess whether an improvement in response of the uterine blood flow to orally administered hormone therapy resulted in an increase in the probability of pregnancy in women with a poor perfusion response in the mid-secretory phase. Animal studies have suggested that older animals are less fertile because of decreased uterine perfusions[37].

A decrease in the collateral circulation of the ovaries from anastomotic channels in women who have undergone abdominal hysterectomy results in earlier menopause[38]. Jansen and Jansson[39] showed a reduction of 52–89% in ovarian blood flow due to hysterectomy. These studies suggested that women may be infertile because of decreased uterine perfusion caused by aging and pelvic surgery[35, 39]. Taylor and colleagues have demonstrated changes in waveforms of ovarian, uterine and iliac vessels associated with changes in the endocrine environment[2].

A modified resistance index demonstrates increasing uterine perfusion with rising estrogen levels in the follicular phase, decreasing uterine perfusion with preovulatory flow and rise in estrogen levels and a subsequent increase in uterine perfusion with rising progesterone levels in the luteal phase[40, 41]. Changes in the uterine waveforms associated with hormonal changes can be clearly assessed using transvaginal color Doppler[42]. Increased diastolic color and pulsed Doppler flow would indicate increased uterine perfusion[43].

The first clinical application of blood flow studies in the management of infertility was reported by McSweeney and colleagues[33]. Authors have demonstrated significant differences in blood supply to the corpus luteum between conception and non-conception cycles after *in vitro* fertilization (IVF). Using the same technique, Deutinger and colleagues[16] have observed differences in blood flow in the ovarian arteries and a correlation between blood flow and hormonal changes in patients undergoing IVF and embryo transfer. An increased blood flow to the ovary containing the preovulatory follicle was reported by Taylor and colleagues[2]. A significantly lower impedance on the side of the developing follicle is present during the luteal phase of the cycle.

The color Doppler signal from ovarian arteries can be identified laterally and below the ovaries. Transvaginal color Doppler investigation of the ovarian arteries reveals additional information to that gained from hormonal parameters and follicular measurements. It can be correlated with the cycle days and the corresponding hormonal profile[16]. A remarkable change of ovarian arterial compliance during the menstrual cycle is noted. In the active ovary, diastolic Doppler shifts increase slightly in the late follicular phase and a relatively high diastolic flow and low pulsatility, consistent with vasodilatation, is seen in the early and late luteal phase. In the inactive ovary, the waveform revealed no diastolic flow, and there appeared to be a high distal vascular impedance throughout the menstrual cycle[43, 44]. There are some reports on luteal flow in domestic and farm animals[45–51]. Various non-invasive and invasive techniques[45, 46] have shown that 80–90% of the blood flow to the ovary is supplied to the corpus luteum. In rabbits and sheep, luteal blood flow has been shown to be more than ten times that in surrounding tissue[49, 50]. This high blood supply is necessary for the supply of the precursors of steroidogenesis and for the removal of synthesized progesterone. A reduction of 90% in luteal flow accompanying luteolysis has been reported following administration of prostaglandin $F_{2\alpha}$[46, 50, 51].

Transvaginal color Doppler can be used in the detection of possible causes of infertility and in the treatment of infertile couples.

Congenital uterine anomalies or uterine fibroids may be the causative factor for infertility or habitual abortions. Sometimes color is detected in the border of the myoma, but the resistance index is always more than 0.50[40, 41].

Hydrosalpinx or sactosalpinx may be detected as an elongated mass filled with fluid or, on transverse sections, as a cystic formation with internal indentations[52]. A tubo-ovarian abscess consists of a dilated, tortuous fluid-filled lumen of the tube, and a solid 'ovarian' part which loses its normal structure of stroma and follicles. The transvaginal probe can also be used in order to assess tenderness and mobility and this can significantly assist in the differential diagnosis[53, 54]. Pelvic inflammatory disease usually shows an absence of detectable blood flow.

Transvaginal sonography may strongly support the clinical suspicion of endometriosis. The 'chocolate cyst', which can be single (Figure 4.17) or multiple (Figure 4.18), unilateral or bilateral, appears as a non-echogenic cyst with a semi-solid content, homogenic or partially cystic[52]. In all examined cases of endometrial cysts in infertile women, a color signal could not be seen inside the mass.

By using ultrasound, it is possible to evaluate the number of developing follicles in the treatment cycle and to improve the timing of follicular puncture. The transvaginal ultrasonically guided puncture represents the most successful and convenient way of collecting oocytes for *in vitro* fertilizations[54, 55]. Transvaginal color Doppler can be used in an *in vitro* fertilization program in order to evaluate the blood flow pattern and correlate it to the overall success rate of IVF therapy. The distance traversed by the needle during the egg retrieval procedure is shorter[54, 55] and the use of Doppler color practically avoids blood vessel injury.

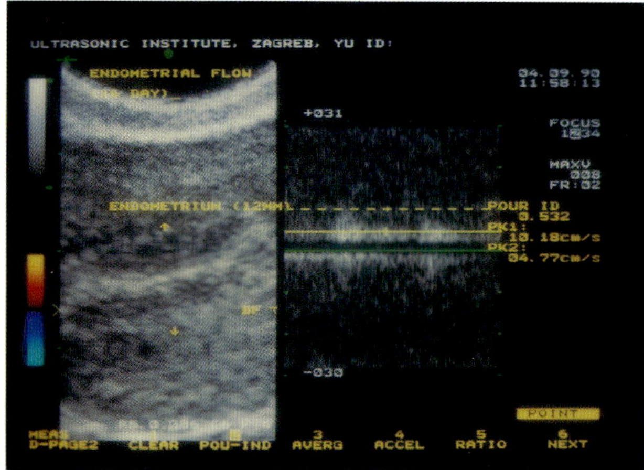

Figure 4.13 Doppler signals obtained from the 12 mm thick endometrium on the 16th day of the cycle. The typical appearance of the preovulatory endometrium exhibits increased basal layer echogenicity (left). A low-impedance flow velocity waveform was obtained from the endometrium (right)

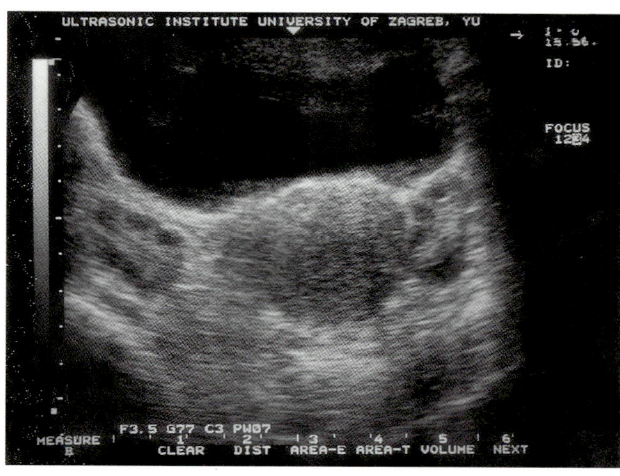

Figure 4.14 Moderately enlarged and polycystic ovaries in a patient suffering from primary infertility

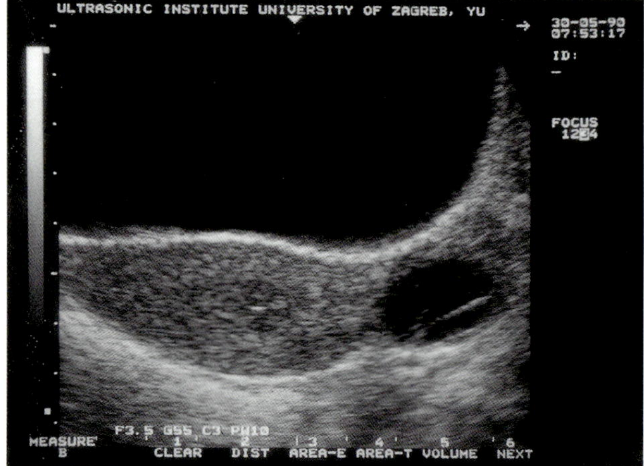

Figure 4.15 A follicular cyst in the left ovary. The patient was referred for monitoring and the follicle did not rupture. Normal ovarian parenchyma is visible on the upper pole of the cyst

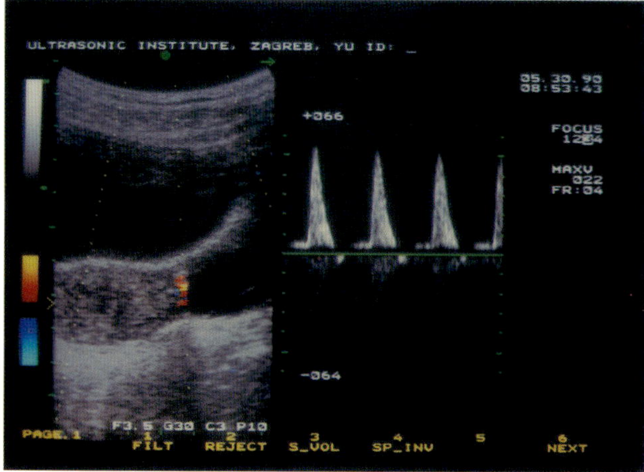

Figure 4.16 Demonstration of blood flow in the uterine artery in an infertility patient (left), showing no diastolic flow (right)

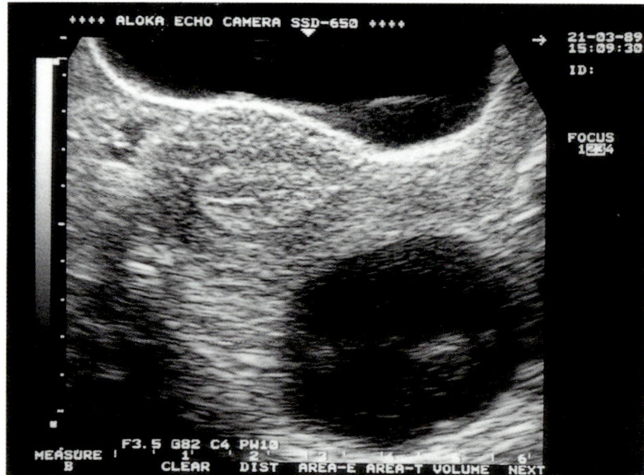

Figure 4.17 Large endometrioma on the left side with pronounced internal echoes

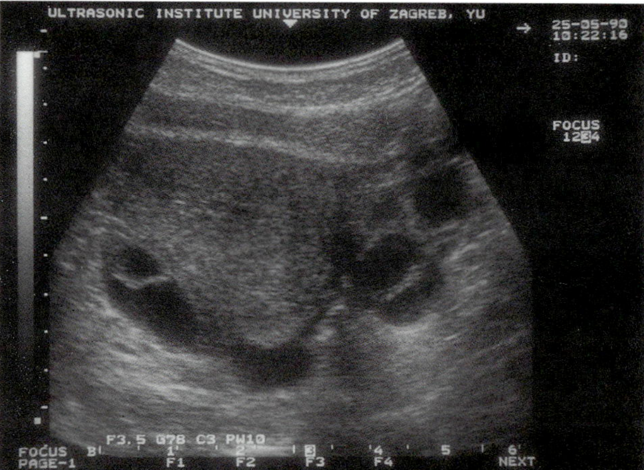

Figure 4.18 The transverse scan showing both ovaries filled with cysts of different sizes and an irregular cyst in the retro-uterine region

CONCLUSION

Ultrasound is one of the most important techniques for the study of female infertility because of its low cost, non-invasiveness and safety. The introduction of transvaginal sonography opens exciting avenues in the investigation of physiological changes as well as pelvic pathology. Ultrasound examination of the pelvic circulation by the transvaginal probe, combined with pulsed and color Doppler assessment, simplifies the examination technique and increases the reproducibility of measurements. An important part of the investigation of reproductive disorders is the examination of small vascular branches and this is facilitated by the use of color Doppler. By this sensitive method, it is possible to differentiate follicular as well as endometrial flow. Further work is necessary to confirm transvaginal color imaging as a part of routine work in the management of infertility.

REFERENCES

1. Kurjak, A. and Jurković, D. (1989). Ultrasonic monitoring of follicular growth and ovulation in spontaneous and stimulated cycles. In Kurjak, A. (ed.) *Ultrasound and Infertility*, p. 90. (Boca Raton, Florida: CRC Press)
2. Taylor, K.J.W., Burns, P.N.T., Conway, D.L. and Hull, M.G.K. (1985). Ultrasound Doppler flow studies of the ovarian and uterine arteries. *Br. J. Obstet. Gynaecol.*, **92**, 240
3. Kurjak, A. and Jurković, D. (1986). New ultrasonic techniques for assessing circulation in the female pelvis. In Kurjak, A. and Kossoff, G. (eds.) *Recent Advances in Ultrasound Diagnosis*, Vol. 5, p. 198. (Amsterdam: Excerpta Medica)
4. Lenz, S. (1985). Ultrasonic study of follicular maturation, ovulation and development of the corpus luteum during normal menstrual cycles. *Acta Obstet. Gynecol. Scand.*, **64**, 15
5. Funduk-Kurjak, B. and Kurjak, A. (1982). Ultrasound monitoring of follicular maturation and ovulation in normal menstrual cycles and ovulation induction. *Acta Obstet. Gynecol. Scand.*, **61**, 329
6. Jansen, C.A.M. (1989). Follicular and endometrial development in the normal ovulatory cycle. Biennial *Erasmus Symposium in Obstetric and Gynecological Ultrasound*, Rotterdam, November 2–4, p. 70 (abstr.)
7. Baber, R.J., McSweeney, M.B., Gill, R.W., Porter, R.N., Picker, R.H., Waren, P.S., Kossoff, G. and Saunders, D.M. (1988). Transvaginal pulsed Doppler ultrasound assessment of blood flow to the corpus luteum in IVF patients following embryo transfer. *Br. J. Obstet. Gynaecol.*, **95**, 1226
8. Baird, D.T. and Fraser, I.S. (1974). Blood production and ovarian secretion rate of estradiol and estrone in women throughout the menstrual cycle. *J. Clin. Endocrinol. Metab.*, **38**, 1009
9. Hackeloer, B.J., Fleming, R., Robinson, H.D., Adam, A.H. and Coutis, J.R.T. (1979). Correlation of ultrasonic and endocrinologic parameters of human follicular development. *Am. J. Obstet. Gynecol.*, **135**, 122
10. Bomsel-Helmreich, O., Huyen, L., Durand-Gasselin, I., Salat-Baroux, J. and Antoine, J.M. (1987). Timing of nuclear maturation and cumulus dissociation in human oocytes stimulated with clomiphene citrate, human menopausal gonadotropin. *Fertil. Steril.*, **48**, 586
11. Almstrong, D.T., Grinwich, D.L., Moon, Y.S. and Zamechnik, J. (1974). Inhibition of ovulation in rabbits by intrafollicular injection of indomethacin and prostaglandin F antiserum. *Life Sci.*, **14**, 129
12. De Crespigny, L., O'Herlihy, C. and Robinson, H.P. (1981). Ultrasonic observation of the mechanism of human ovulation. *Am. J. Obstet. Gynecol.*, **139**, 177
13. Kerin, J.F., Edmonds, D.K., Warnes, G.M., Cox, L.W., Seamark, R.F., Matthews, C.D., Young, G.B. and Baird, D.T. (1981). Morphological and functional relationship of Graafian follicle growth to ovulation in women using ultrasonic, laparoscopic and biochemical measurements. *Br. J. Obstet. Gynaecol.*, **88**, 81
14. Picker, R.H., Smith, D.H., Tucker, M.H. and Saunders, D.M. (1983). Ultrasonic signs of imminent ovulation. *J. Clin. Ultrasound*, **11**, 1
15. Davis, F.A. and Gosink, B.B. (1986). Fluid in the female pelvis. Cyclic patterns. *J. Ultrasound Med.*, **5**, 75
16. Deutinger, J., Reinthaller, A. and Bernaschek, D. (1989). Transvaginal pulsed Doppler measurement of blood flow velocity in the ovarian arteries during cycle stimulation and after follicle puncture. *Fertil. Steril.*, **54**, 466
17. Blumenfeld, Z. and Nahhas, F. (1988). Luteal dysfunction in ovulation induction; the role of repetitive chorionic supplementation during the luteal phase. *Fertil. Steril.*, **50**, 403
18. Fleischer, A., Gordon, N.A., Entman, S.S. and Kepple, D.M. (1990). Transvaginal scanning of the endometrium. *J. Clin. Ultrasound*, **18**, 337
19. Sakamoto, C. and Nakano, H. (1982). The echogenic endometrium and alterations during the menstrual cycle. *Int. J. Gynaecol. Obstet.*, **20**, 255
20. Christie, A.D. (1981). *Ultrasound and Infertility*, p. 19. (Bromley: Chartwel-Bratt)
21. Itskovitz, J., Bolders, R., Levron, J. and Thaler, I. (1990). Transvaginal sonography in the diagnosis and treatment of infertility. *J. Clin. Ultrasound*, **18**, 249
22. Adams, J.M., Tan, S.L., Wheeler, M.J., Morris, D.L., Jacobs, H.S. and Franks, S. (1988). Uterine growth in the follicular phase of spontaneous ovulatory cycles and during luteinizing hormone-releasing hormone-induced cycles and during luteinizing hormone-induced cycles in women with normal or polycystic ovaries. *Fertil. Steril.*, **49**, 52
23. Glissant, A., de Mouzon, J. and Frydman, R. (1985). Ultrasound study of the endometrium during *in vitro* fertilization cycles. *Fertil. Steril.*, **44**, 786
24. Fleischer, A.C., Herbert, C.M., Sacks, G.A. *et al.* (1986). Sonography of the endometrium during conception and nonconception cycles for *in vitro* fertilization and embryo transfer. *Fertil. Steril.*, **46**, 786
25. Imoedemhe, D.A.G., Hackeloer, B.J. and Daume, E. (1986). Ultrasound measurement of endometrial thickness on different ovarian stimulation regimens during *in-vitro* fertilization. *Hum. Reprod.*, **2**, 545
26. Rabinowitz, R., Laufer, N., Lewin, A. *et al.* (1986). The value of ultrasonographic endometrial measurement in the prediction of pregnancy following *in vitro* fertilization. *Fertil. Steril.*, **45**, 824

27. Timor-Tritsch, I. and Rottem, S. (1989). Transvaginal sonography in the management of infertility. In Kurjak, A. (ed.) *Ultrasound and Infertility*, p. 126. (Boca Raton, Florida: CRC Press)
28. Ross, G.T. and Schreiber, J.R. (1986). The ovary. In Yen, S.S.C. and Jaffe, R.B. (eds.) *Reproductive Endocrinology*, p. 115. (Philadelphia: W.B. Saunders)
29. Jewelewicz, R. (1975). Management of infertility resulting from anovulation. *Am. J. Obstet. Gynecol.*, **122**, 309
30. Coulam, C.B., Hill, L.M. and Breckle, R. (1982). Ultrasonic evidence for luteinization of unruptured preovulatory follicle. *Fertil. Steril.*, **37**, 524
31. Hamilton, C.J.C.M., Evers, J.L.H. and Hoogland, H.J. (1986). Ovulatory disorders and inflammatory adnexal damage: a neglected cause of the failure of fertility microsurgery. *Br. J. Obstet. Gynaecol.*, **93**, 282
32. Thomas, E.J., Lenton E.A. and Cooke, I.D. (1986). Follicle growth patterns and endocrinological abnormalities in infertile women with minor degrees of endometriosis. *Br. J. Obstet. Gynaecol.*, **93**, 852
33. McSweeney, M.B., Baber, R.J., Gill, R.W., Kossoff, G., Saunders, D., Porter, R. and Picker, R. (1988). Prediction of IVF and GIFT outcome using transvaginal Doppler assessment of ovarian blood flow. *J. Ultrasound Med.*, 7, 73
34. Kurjak, A., Žalud, I. and Crvenković, G. (1989). Transvaginal color Doppler in the assessment of maternal and fetal circulation. *Biennal Erasmus Symposium in Obstetric and Gynecological Ultrasound*, Rotterdam, November 2–4, p. 93 (abstr.)
35. Goswamy, R.K., Williams, G. and Steptoe, P.C. (1988). Decreased uterine perfusion – a cause of infertility. *Hum. Reprod.*, **3**, 955
36. Goswamy, R.K. and Steptoe, P.C. (1988). Doppler ultrasound studies of the uterine artery in spontaneous ovarian cycles. *Hum. Reprod.*, **3**, 721
37. Finch, C.E. and Gosden, R.M. (1986). Animal models for the human menopause. In Mastroiani, L.I. and Paulsen, C.A. (eds.) *Ageing, Reproduction and the Climacteric*, p. 3. (New York: Plenum Press)
38. Siddle, N., Sarrel, P. and Whitehead, M.I. (1987). The effect of hysterectomy on the age of ovarian failure: identification of a subgroup of women with premature loss of ovarian function and literature review. *Fertil. Steril.*, **47**, 94
39. Jansen, P.O. and Jansson, I. (1977). The acute effect of hysterectomy on ovarian blood flow. *Am. J. Obstet. Gynecol.*, **127**, 349
40. Kurjak, A., Žalud, I., Alfirevic, Z. and Jurković, D. (1990). The assessment of abnormal pelvic blood flow by transvaginal color Doppler and pulsed Doppler. *Ultrasound Med. Biol.*, in press
41. Kurjak, A., Žalud, I., Jurković, D., Alfirevic, Z. and Miljan, M. (1989). Color Doppler in the assessment of pelvic circulation. *Acta Obstet. Gynecol. Scand.*, **68**, 131
42. Kurjak, A., Jurković, D., Alfirevic, Z. and Žalud, I. (1990). Transvaginal color Doppler imaging. *J. Clin. Ultrasound*, **18**, 227
43. Hata, K., Hata, T., Senoh, D., Makihara, K., Aoki, S., Takamiya, O. and Kitao, O. (1990). Change in ovarian arterial compliance during the human menstrual cycle assessed by Doppler ultrasound. *Br. J. Obstet. Gynaecol.*, **97**, 163
44. Žalud, I. and Kurjak, A. (1990). The assessment of luteal blood flow in pregnant and non-pregnant women by transvaginal color Doppler. *J. Perinat. Med.*, **18**, 215
45. Abdul-Karim, R.W. and Bruce, N.W. (1973). Blood flow to the ovary and corpus luteum at different stages of gestation in the rabbit. *Fertil. Steril.*, **24**, 44
46. Bruce, N.W. and Hiller, K. (1974). The effect of prostaglandin F on ovarian blood flow and corpora lutea regression in the rabbit. *Nature (London)*, **249**, 176
47. Bruce, N.W. and Meyer, G.T. (1981). Ovarian blood flow and progesterone secretion in anesthetized rats at day 16 of gestation, and the effects of hemorrhage. *J. Reprod. Fertil.*, **61**, 419
48. Bruce, N.W. and Moor, R.M. (1975). Ovarian blood flow and function. *Bibliogr. Reprod.*, **25**, 581
49. Bruce, N.W. and Moor, T.J. (1976). Capillary blood flow to ovarian follicles, stroma and corpora lutea of anesthetized sheep. *J. Reprod. Fertil.*, **46**, 29
50. Novy, M.J. and Cook, M.J. (1973). Redistribution of blood flow by prostaglandin $F_{2\alpha}$ in the rabbit ovary. *Am. J. Obstet. Gynecol.*, **117**, 381
51. Nett, T.M. and Niswender, G.D. (1981). Luteal blood flow and receptors for LH during $PGF_{2\alpha}$-induced luteolysis: production of $PGF_{2\alpha}$ during early pregnancy. *Acta Vet. Scand.* (Suppl.), 77, 117
52. Timor-Tritsch, I. and Rottem, S. (1989). Transvaginal sonography in the management of infertility. In Kurjak, A. (ed.) *Ultrasound and Infertility*, p. 126. (Boca Raton, Florida: CRC Press)
53. Timor-Tritsch, I. and Rottem, S. (1987). *Transvaginal Sonography*, p. 143. (New York: Elsevier Science)
54. Feichtinger, W. and Kemeter, P. (1986). Transvaginal sector scan sonography for needle aspiration and other applications in gynecologic routine and research. *Fertil. Steril.*, **45**, 722
55. Dellenbach, P., Nisand, I., Moreau, L., Feger, B., Plumere, C. and Gerlinger, P. (1985). Transvaginal sonographically controlled follicle puncture for oocyte retrieval. *Fertil. Steril.*, **44**, 656

5 Assessment of Early Pregnancy

A. Kurjak, G. Crvenković and H. Takeuchi

More than any other methods available so far, ultrasound has enabled us to obtain direct information on embryo differentiation and growth[1]. The embryonic period is probably the most critical period of intrauterine life, characterized by many changes which will influence the development of pregnancy in the later stages.

Remarkable progress in the evaluation of this period of pregnancy has been obtained by means of transvaginal color Doppler. This new technique of investigation offers a series of new physiological data, but also enables us to observe *in utero* the features classically described by the embryologist. Transvaginal color Doppler combined with high-resolution ultrasound allows a very close examination of the fine structures and the developmental events occurring in embryonic life. Although the transvaginal probe (5 MHz) is in close contact with the developing tissues, the spatial-peak temporal average intensity of the equipment used (Aloka Color Doppler SSD-680) is approximately 80 mW/cm^2.

This chapter summarizes our experience with color Doppler investigation of maternal–embryonal hemodynamics in the first trimester of normal and pathological pregnancies.

THE 4TH WEEK

Main characteristic embryonical findings[2]

The deep neural groove and the first somites are present. The embryo is almost straight and the somites produce conspicuous surface elevation. The heart prominence is distinct and the optic pits are present. At the end of the 4th week, the attenuated tail with its somites is also a characteristic feature.

Ultrasound findings

At the 4th week and 1 day of menstrual age the gestational sac, which is 5 mm in diameter, can be detected by transvaginal ultrasonography as a tiny ring-shaped structure[3] (Figure 5.1). It is important to point out here that, by means of ultrasound methods, it may not always be easy to differentiate gestational sacs from other ring-like structures, particularly the 'pseudogestational sac' of ectopic pregnancy at this gestational age[3].

Transvaginal color Doppler findings

Blood flow in intervillous spaces can be observed by transvaginal color Doppler as early as 4 weeks and 4 days from the last menstrual period (Figure 5.2). A color signal from these vessels is seen within the hyperechoic area in close proximity to the gestational sac. The blood flow velocities in systole and diastole are very low (Figure 5.3). A clear and accurate display of the gestational ring and the presence of intervillous space flow on its periphery are suggestive of a normal intrauterine pregnancy.

The Doppler measurements at the level of the uterine arteries are not essentially different from the findings in non-gravid women[4].

THE 5TH WEEK

Main characteristic embryonical findings[2]

The embryo has a C-shaped curve. The growth of the head (caused by the rapid development of the brain) exceeds that of the other regions.

Ultrasound findings[3]

The gestational sac is always visible and is surrounded by a broad, echodense asymmetric ring. The gestational sac is therefore easily measurable, but a prognosis cannot yet be made as to the outcome of the pregnancy.

The secondary yolk sac can be identified, but the embryonic pole is not always discernible at the beginning of this week (Figure 5.4).

Using the vaginal probe, the embryo length can be measured and varies from 3 to 5 mm (Figure 5.5).

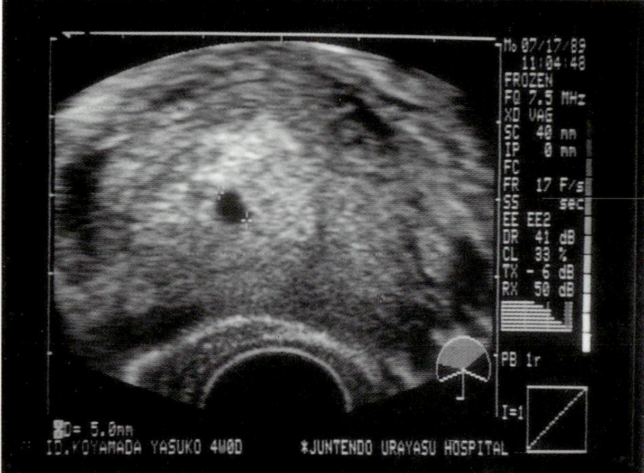

Figure 5.1 This echogram shows a small gestational sac of diameter 5 mm at 17th day after ovulation. Echogenic ring as gestational sac with surrounding mass of decidua can be observed. No particular structure can be seen inside of gestational sac

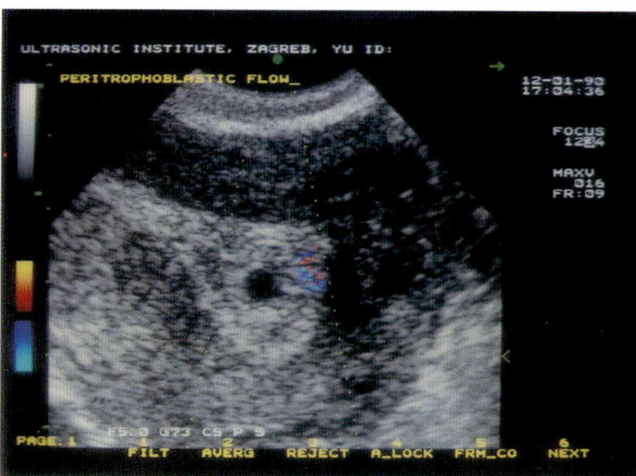

Figure 5.2 Vascularization through the intervillous spaces by color Doppler in a normal 4-week-old pregnancy

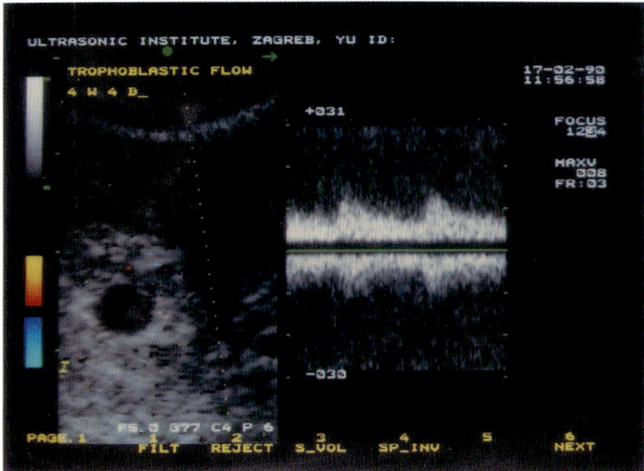

Figure 5.3 Low-impedance flow velocity waveforms (right) that were obtained from the intervillous space vessels at 4 weeks and 4 days of gestation

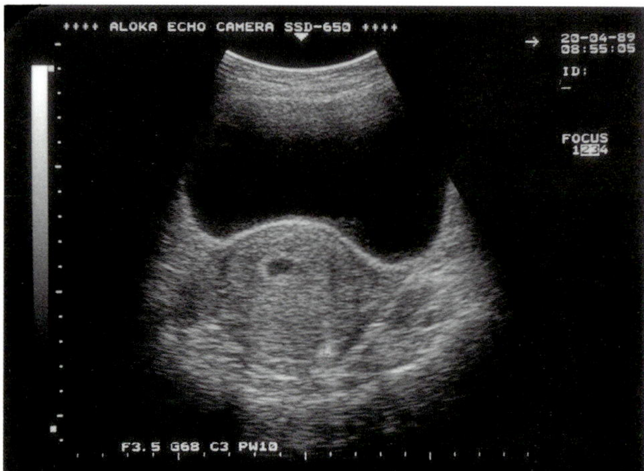

Figure 5.4 A normal 5-week 2-day pregnancy. Note the yolk sac within the gestational sac

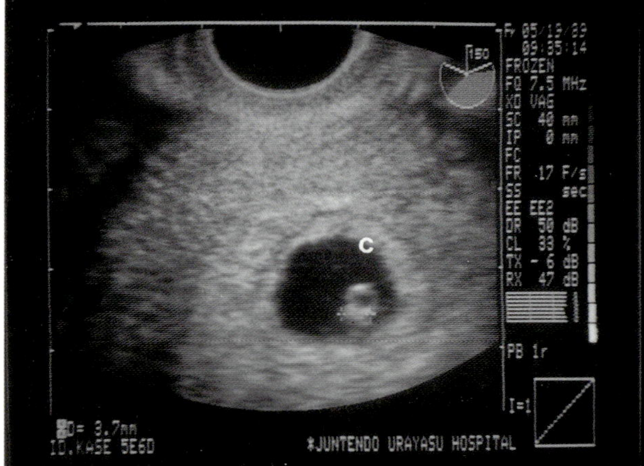

Figure 5.5 This echogram shows an embryonic pole of maximum length 3.7 mm, which corresponds to 5 weeks plus 5 days of pregnancy. At this stage, no fine structure of an embryo can be differentiated. However, the chorion has almost equal thickness for 4–5 weeks of pregnancy; a part of this chorion (c) shows the decrease in thickness. This is the regression of the chorion, known as chorion leave

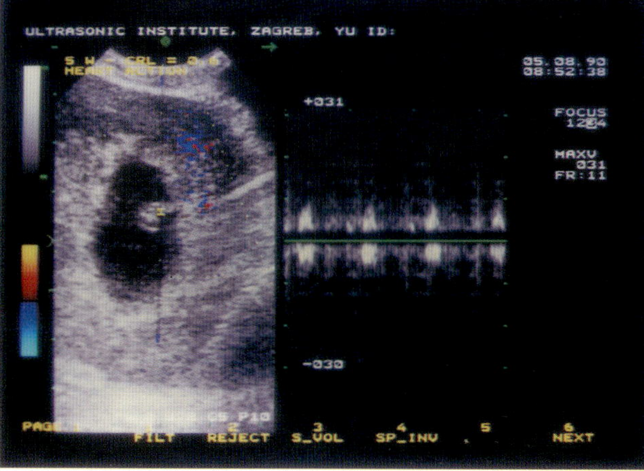

Figure 5.6 Color Doppler images of the embryonal heart activity showing absence of diastolic flow (right) at 5 weeks of gestation

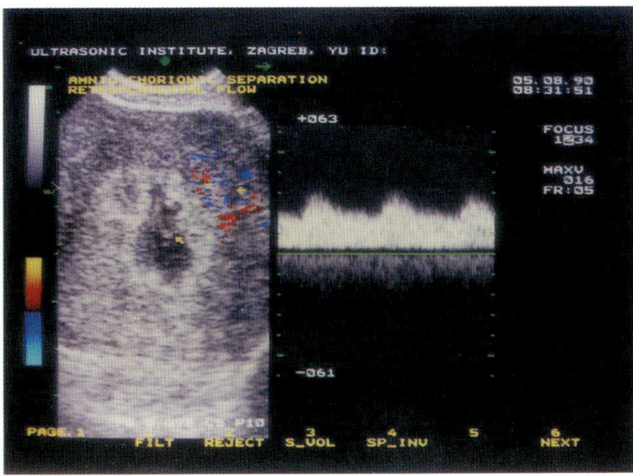

Figure 5.7 Doppler study of the flow through intervillous spaces showing high diastolic flow at 5 weeks

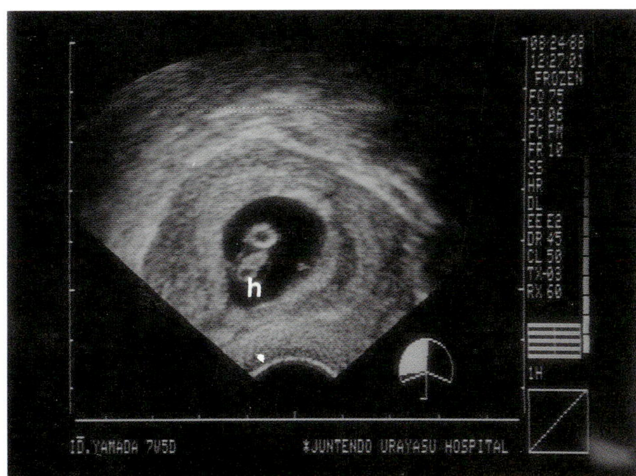

Figure 5.8 In a transverse section of a 5-mm crown–rump length embryo, corresponding to 6 weeks and 0 days of pregnancy, an apparent anechoic cavity (h) is seen in the front part of the head. This cavity shows the forebrain (prosencephalon). Amnion is visualized close to the embryo. The yolk sac is also depicted very clearly

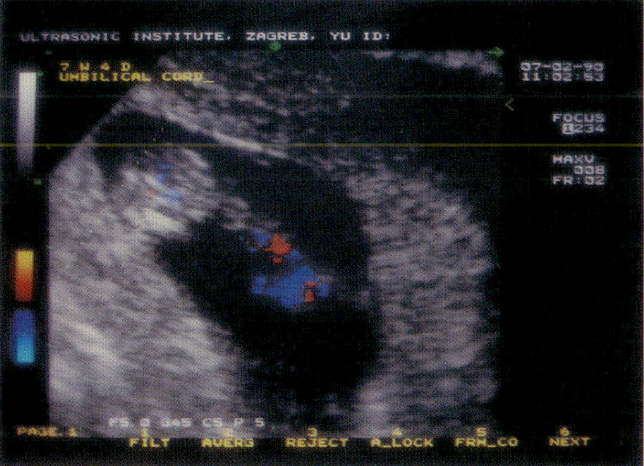

Figure 5.9 Blood flow in the umbilical arteries as seen by color Doppler at 7 weeks and 4 days of gestation

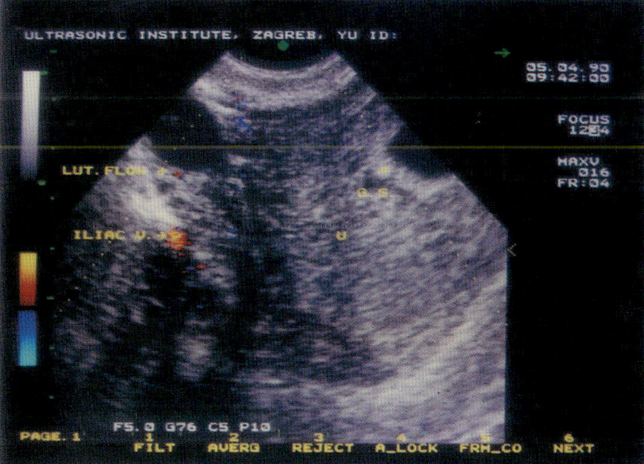

Figure 5.10 Presence of luteal flow close to the cyst through one of the ovaries during early pregnancy. Iliac vessels are also clearly visible. G.S. = gestational sac; U = uterus

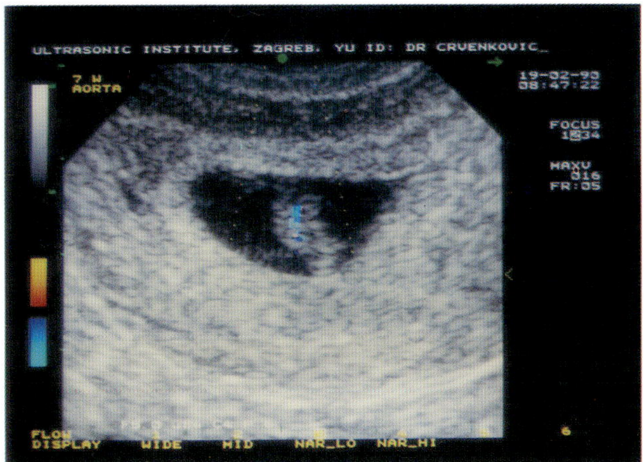

Figure 5.11 Blood flow through the fetal aorta at the 7th week of gestation

Transvaginal sonography is able to detect and clearly shows lacunar formations at one pole of the implanted gestation[5]. Embryonic structures are usually not identifiable with ultrasound at this stage. Embryonal heart beats can be detected by means of sonography at 5 weeks and 3 days of menstrual age at the earliest.

Transvaginal color Doppler findings

Color accurately indicates the position of embryonal heart pulsations (Figure 5.6). At this very early stage this finding may help clinicians to differentiate, in cases of missed abortion, between a living embryo and a non-living embryo.

The values of the resistance index in uterine arteries do not show a significant difference in relation to the flow through such blood vessels in non-gravid women[6].

A luteal flow through one of the ovaries can be observed easily at this gestational age.

In the 5th week of pregnancy, blood flow in the intervillous spaces can be visualized clearly in every case (Figure 5.7).

THE 6TH WEEK

Main characteristic embryonical findings[2]

The head is now much larger in relation to the trunk and is more bent over the cardiac prominence. The trunk and neck have begun to straighten. Hand and foot plates are formed. Digital or finger rays have appeared.

Ultrasound findings

The gestational sac occupies about one third of the uterine volume[7]. Between 6 and 7 weeks the embryo passes the 5 mm limit, and thus the crown–rump length can now be measured reliably (Figure 5.8). At this gestational period, an adnexal or luteal cyst can be clearly detected.

Transvaginal color Doppler findings

A narrow uterine flow is noted in 73% of pregnant women. Compared to the 5th week of pregnancy, the value of the resistance index through the uterine arteries increases slightly.

THE 7TH WEEK

Main characteristic embryonical findings[2]

By the beginning of the 7th week of development, the embryo has acquired a skeleton, which is mostly cartilaginous and which gives form to the body of the embryo.

The communication between the primitive gut and the yolk sac has been reduced to a relatively small duct (the yolk stalk).

Ultrasound findings

By the 7th week it should be possible to see an 'embryonal pole' as a little blob within the gestational sac[8]. Fetal movements seldom occur at 7 weeks of menstrual age[9].

Transvaginal color Doppler findings

A flow through the umbilical artery, characterized by the absence of the flow in diastole and a resistance index equal to 1, can be detected at this stage of gestation (Figure 5.9). This finding may be explained by the slow, low-volume umbilical cord flow at this gestational age which cannot be detected with the present color Doppler sensitivity and 100 Hz high-pass-filter in pulsed wave Doppler[10].

The mean values of the resistance index through uterine arteries are identical to the measurements observed at earlier gestational ages.

The luteal flow is detectable in 80% of pregnant women in this period (Figure 5.10).

The aorta can be visualized using transvaginal color Doppler starting from 7 weeks of gestation (Figure 5.11). In the umbilical cord, waveform analysis shows no diastolic flow and a high, regular systolic Doppler shift.

THE 8TH–9TH WEEKS

Main characteristic embryonical findings

The head is more rounded and constitutes almost half of the embryo[2]. The hands and feet approach each other. Fingers are free and longer. The intestines are in the umbilical cord (physiological hernia)[11].

Ultrasound findings[3]

The fetal structures are now clearly discernible (Figures 5.12 and 5.13), in particular the physiological hernia is easily visible. The yolk sac frequently appears as a circular structure of 5 mm diameter. The yolk sac must not be mistaken for a twin gestation, and its presence precludes the diagnosis of a blighted ovum[12].

The fetal kidneys and adrenal glands can be imaged as early as 9 menstrual weeks of gestation, when their size is within the resolution of today's equipment[13].

The placenta becomes more demarcated and its relationship to the uterine cavity may be extrapolated[5].

Transvaginal color Doppler findings

No more significant shifts are found compared to previously described values of maternal–fetal hemodynamics, apart from the presence of accelerated blood flow through the blood vessels of the corpus luteum in relation to previous measurements in earlier stages of pregnancy.

THE 10TH WEEK

Main characteristic embryonical findings[2]

The intestines are now back in the abdomen. Urine starts to form and is excreted into the amniotic fluid.

Ultrasound findings[3]

With increasing differentiation of the fetus and rapid progression of ossification, fetal biometry becomes easier at this stage. For example, the biparietal diameter (BPD) can be measured from 8 weeks onwards. For practical applications it is best to use both CRL and BPD measurements (Figure 5.14).

Transvaginal color Doppler findings

At this stage, blood flow can be detected through the middle cerebral artery and fetal heart by transvaginal color Doppler sonography (Figure 5.15).

A notable diastolic flow of small velocity is significant in the middle cerebral artery at 10 weeks of gestation (RI = 1) (Figure 5.16).

All other fetal and maternal parameters measurable by this transvaginal color Doppler (umbilical artery, aorta, uterine arteries, retroplacental flow and corpus luteum flow) (Figure 5.17) do not show any significant changes.

THE 11TH–12TH WEEKS

Main characteristic embryonical findings[2]

Fetal sex is clearly distinguishable by 12 weeks. The neck is well-defined, the face is broad, the eyes are widely separated. By the end of the 12th week, erythropoiesis decreases in the liver and begins in the spleen. The decidua capsularis adheres to the decidua parietalis.

Ultrasound findings[8]

At 11 weeks of menstrual age, the sac wall becomes less distinct and its ultrasonic appearance begins to break up.

The umbilical cord is clearly visible (Figure 5.18).

At this time, about 50% of the placenta seems to occupy partly or completely the lower pole of the uterus (Figure 5.19).

By using endovaginal scanning, the heart, liver and stomach in the trunk, as well as the fourth ventricle in the head, can be observed as organ structures (Figure 5.20).

It is possible to show the head, heart, diaphragm, stomach, kidney, urinary bladder, spine and extremities by using endovaginal sonography at this stage of pregnancy (Figure 5.21).

Transvaginal color Doppler findings

This gestational age is characterized by a slight decrease of the RI values through all the structures comprising the materno–fetal circulation. From this period, the umbilical vein flow is visible as a continuous flow. The RI values of the umbilical artery, fetal aorta and middle cerebral artery are still equal to 1. The RI of the arcuate artery is high in non-pregnant patients and in those in early pregnancy[14].

Discrepancies of blood flow velocity between the two uterine arteries are significant for this gestational age[15]. These left/right differences are more pronounced during early gestation and disappear as pregnancy advances. The explanation for this may be that one artery is the primary provider to the placenta and will undergo changes of resistance much earlier than the anteplacental vessels[16].

The flow through the umbilical vein is simple to measure together with the signal from the umbilical artery. The venous signal has a continuous form (Figure 5.22).

All other relevant flow values do not show any differences from those observed in earlier stages of pregnancy.

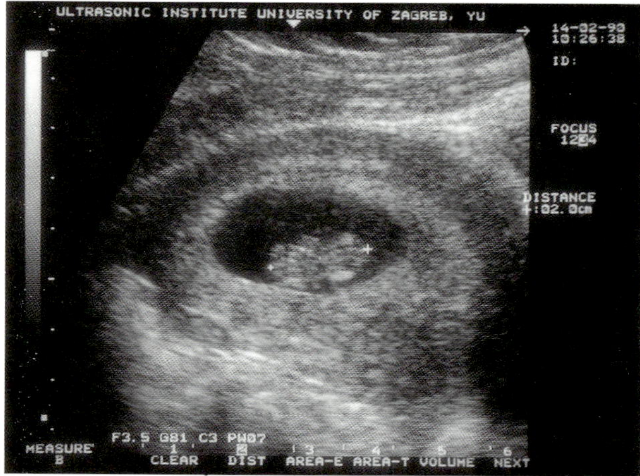

Figure 5.12 An 8-week 5-day pregnancy. CRL = 20 mm. The yolk sac is normal but visible in a different plane

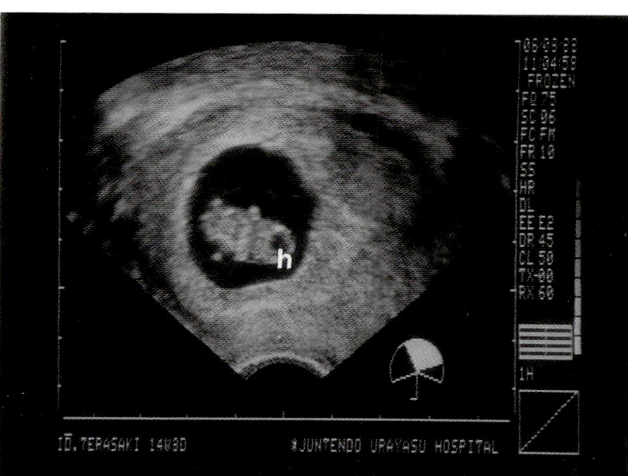

Figure 5.13 In this echogram of 21-mm crown–rump length embryo, the head and the trunk are easily differentiated. The distinct anechoic cavity (h) of the hindbrain is seen in the head. Each of four limb buds is depicted in this coronal section of an embryo

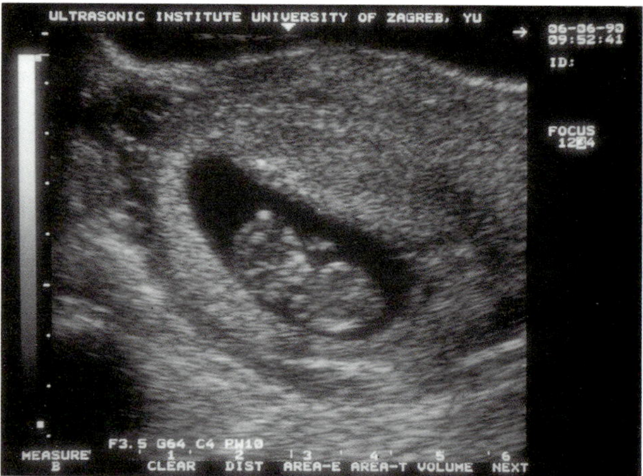

Figure 5.14 Measurement of a 10-week 2-day fetus. CRL = 32 mm

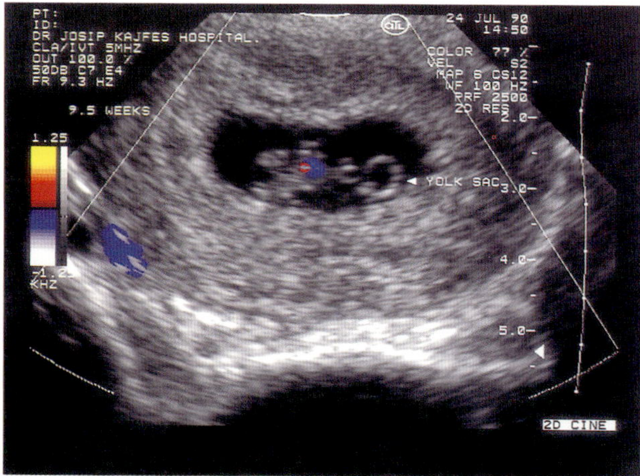

Figure 5.15 Fetal heart visualized by color Doppler. The yolk sac is visible close to the fetus

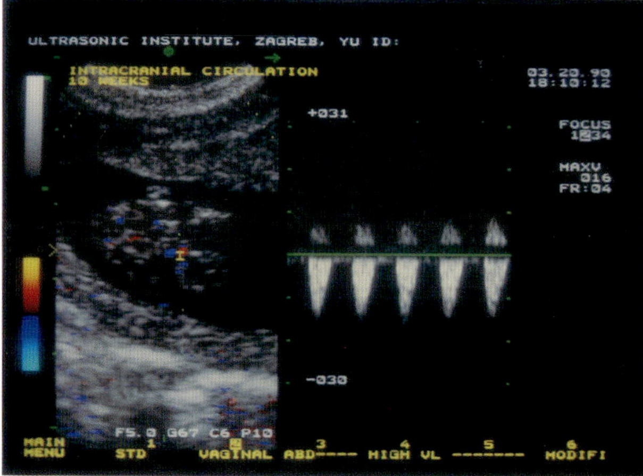

Figure 5.16 Flow through middle cerebral artery (left) and Doppler signal with no diastolic flow at 10 weeks (right)

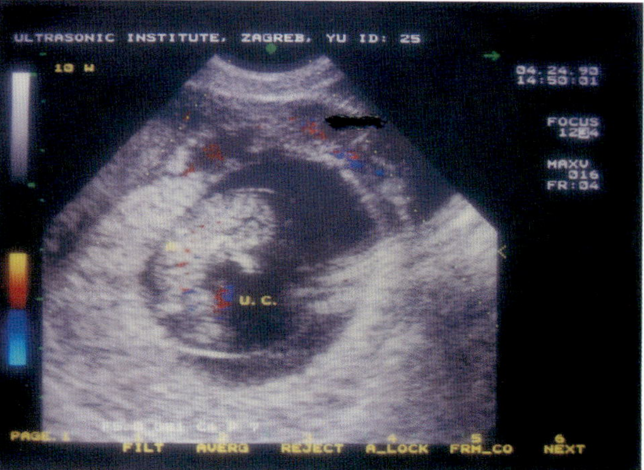

Figure 5.17 Color flow signal of the fetoplacental circulation showing a highly vascularized area through fetal aorta, umbilical cord and retroplacental vessels. A = aorta; U.C. = umbilical cord

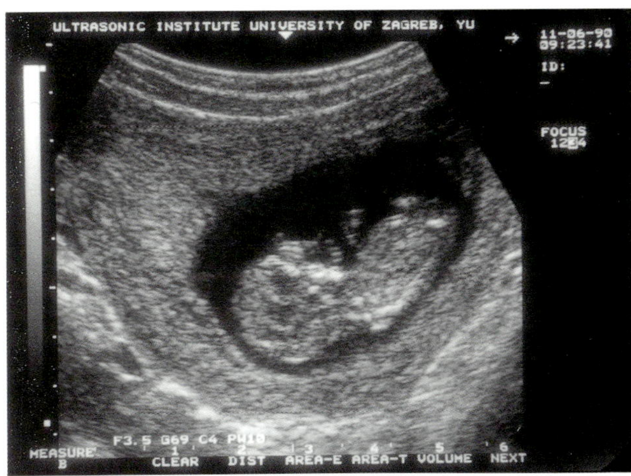

Figure 5.18 Transabdominal scan shows 11-week 2-day fetus with a clearly visible umbilical cord. The yolk sac is visible above the head

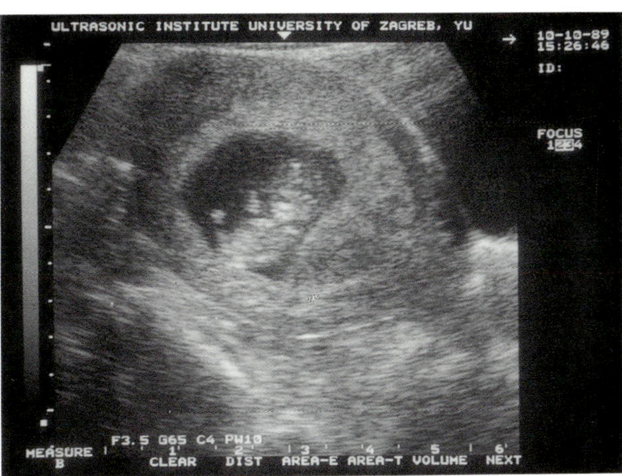

Figure 5.19 Longitudinal scan of an 11-week fetus showing the development of the placenta, partly occupying the lower pole of the uterus

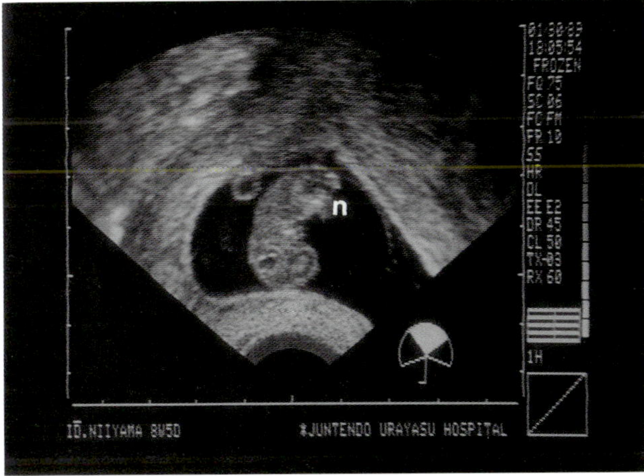

Figure 5.20 In this sagittal section of an embryo of 40-mm crown–rump length, heart, liver and stomach in the trunk, as well as the fourth ventricle in the head, can be observed as organ structures. The stomach is depicted at an unexpectedly lower part of the trunk. The insertion of the umbilical cord (n) into the abdomen is seen larger in this stage of pregnancy because of the physiological herniation of the gut

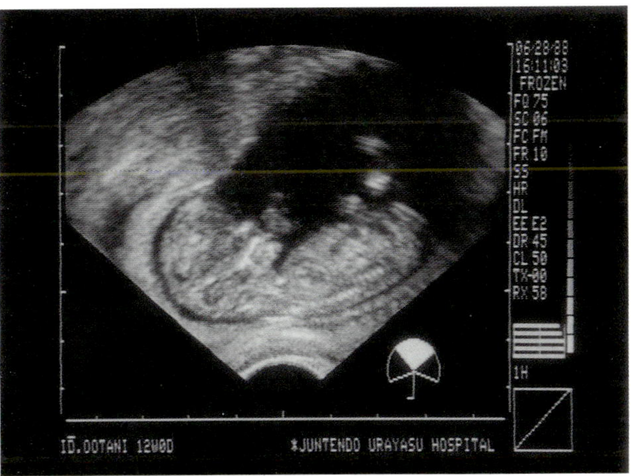

Figure 5.21 This sagittal section of a fetus of 56-mm crown–rump length and 20-mm biparietal diameter (12 weeks plus 3 days of pregnancy) delineates its face very well. Maxilla and mandibular are also clearly shown. The heart in the thorax, and intestine and urinary bladder in the abdomen can be recognized

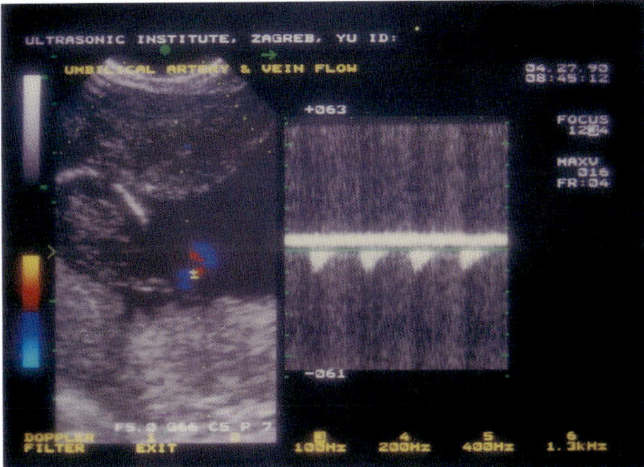

Figure 5.22 Doppler signals obtained from the umbilical artery and vein (left) showing no diastolic arterial flow and continuous venous flow (right) at 11 weeks

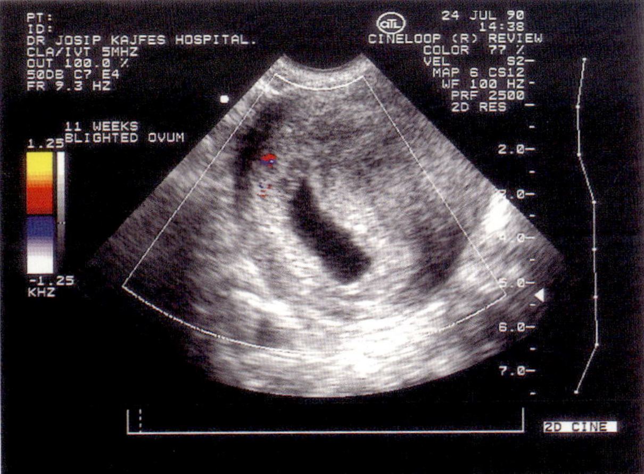

Figure 5.23 Transvaginal scan of a blighted ovum pregnancy shows absence of the embryo, irregular shape and size of gestational sac

The existence of a diastolic flow can be demonstrated later in the second and third trimester.

ABNORMALITIES IN THE FIRST TRIMESTER

The most frequent complication in pregnancy is bleeding and it always raises the question as to whether the pregnancy is normal or abnormal[17]. So far, clinical and laboratory tests have contributed a great amount to the early detection of abnormalities in this phase of pregnancy, but none as yet to such an extent as transvaginal color Doppler sonography. The reasons for this are:

(1) For this kind of examination, it is not necessary for the patient to have a full bladder. This, apart from being comfortable, also has the advantage that the physiological position of the organ is preserved and thus the possibility of relevant structures being out of focus is reduced to a minimum (intestinal localization of ectopic pregnancy);

(2) The time needed for the detection of the present flow is shortened to a greater extent;

(3) A clear display of the flow through the blood vessels is decisive for the final diagnosis (heart and fetal aortae with missed abortion);

(4) A detailed presentation of all pelvic structures due to the closeness of the probe is possible;

(5) An atypical position of the uterus does not hamper the formation of a diagnosis and a detailed analysis of the uterus.

Pathological pregnancy

Blighted ovum

The transvaginal color Doppler method enables one to perform this diagnosis as early as the end of the 5th gestational week. Together with well-known diagnostic criteria detected by conventional ultrasound (i.e. the absence of an embryo; irregular shape and the size of the gestational sac in relation to the gestational age; discrepancy between uterine size and the gestational sac), the absence of the blood flow through the heart of the embryo is characteristic. This fact leads to an urgent diagnosis of such a pathological pregnancy, diminishing the possibility of postabortional complications. Resistance index values of uterine arteries, intervillous space/retroplacental (Figures 5.23, 5.24 and 5.25) and luteal flow are slightly decreased as in the normal intrauterine pregnancy[10].

Missed abortion

It is relatively easy to make this diagnosis by means of the transvaginal color Doppler method. The main parameter is the absence of the heart beat and the lack of a color flow signal at its expected position after the 6th gestational week (Figure 5.26).

The mean resistance index value of the uterine artery flow is lower than the mean value for the same vessels in normal intrauterine pregnancy[10], while the mean resistance index value of retroplacental flow is similar.

The absence of fetal movement is easy to recognize by using this method of ultrasound examination.

Hydatidiform mole

A diagnosis of hydatidiform mole is relatively easy to make from its characteristic 'snowstorm-like' sonographic appearance, which is considered to be pathognomonic for this disorder[3] (Figure 5.27).

The transvaginal approach should be used in obese patients or in patients in whom the image obtained by transabdominal sonography is not conclusive.

Transvaginal color Doppler sonography allows us to register the large volume of flow through small cystic structures inside the uterus (Figure 5.28). The flow through other relevant vessels does not show any difference from that in normal pregnancy.

FETAL MALFORMATION

It is not certain at what gestational age various fetal anomalies can be diagnosed with endovaginal sonography. However, even in the first trimester some anomalies can be found by this method (Figures 5.29 and 5.30). Besides anencephaly and omphalocele, nuchal cystic hygroma is reported as a fetal anomaly detected in the first trimester (Figures 5.31 and 5.32).

CONCLUSION

The mystery of intrauterine life is becoming less and less of a mystery and more and more of a reality[18]. The application of high-frequency transvaginal sonography offers new opportunities in scanning during the first trimester of pregnancy[19]. Transvaginal ultrasound examinations might improve diagnostic accuracy in localizing early pregnancies, as compared to transabdominal techniques[20].

Doppler ultrasound has opened the possibility of

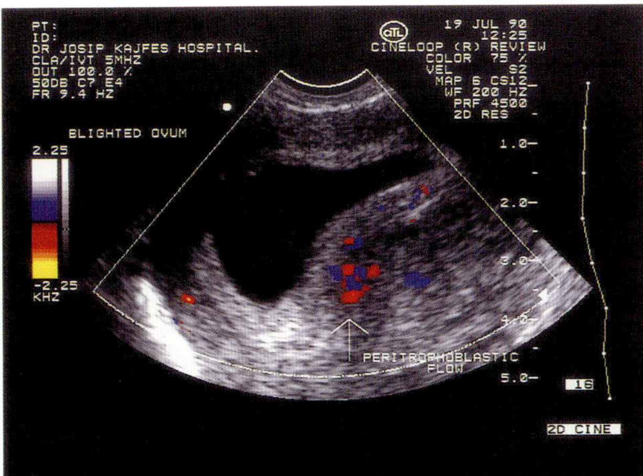

Figure 5.24 Another example of a blighted ovum pregnancy. Peritrophoblastic blood flow was detected by color Doppler

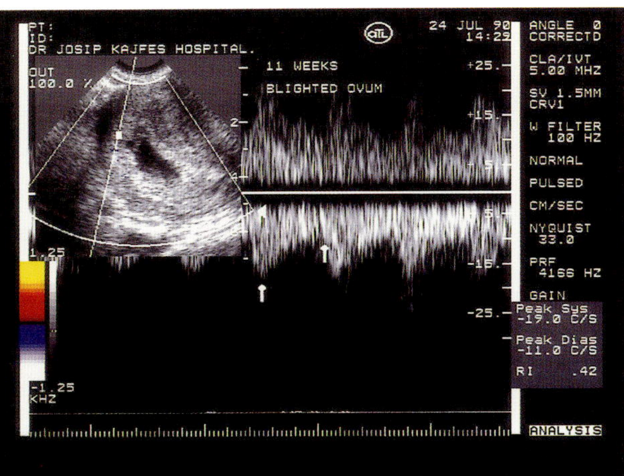

Figure 5.25 Low-impedance flow velocity waveforms (right) that were obtained from the intervillous space vessels in the case of a blighted ovum. Absence of the embryonal pole is evident

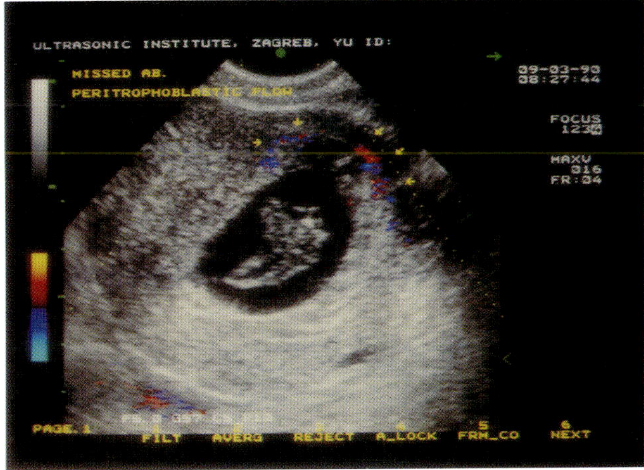

Figure 5.26 Color flow in the case of missed abortion showing a highly vascularized area close to the gestational sac (arrows indicate the intervillous space vessels). No cardiac activity

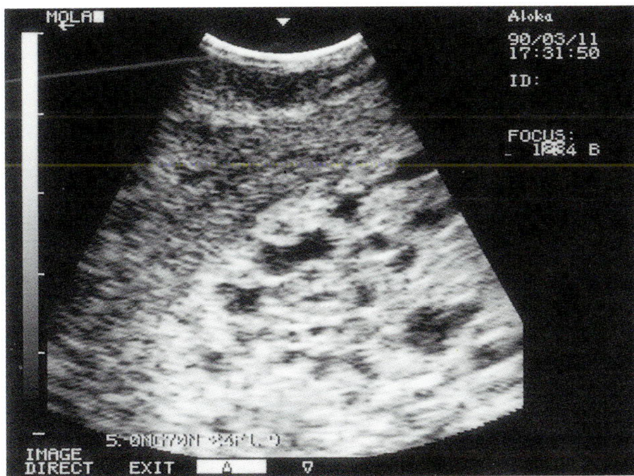

Figure 5.27 Hydatidiform mole. The transvaginal image is typical, showing a large number of sonolucent round structures of varying sizes

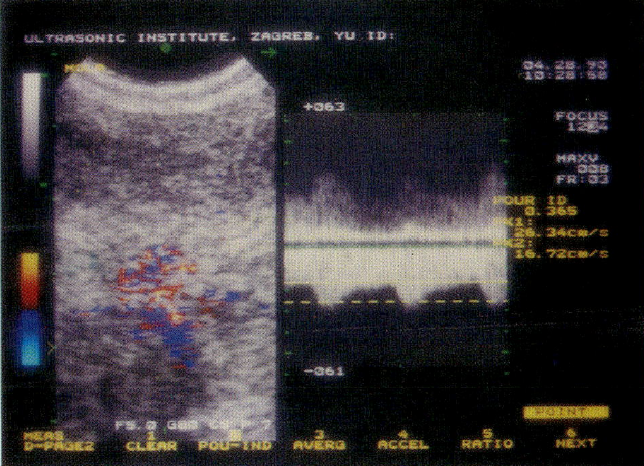

Figure 5.28 Pulsed Doppler signal from the vascular region of the hydatidiform mole indicates the presence of low-impedance flow (right). Increased vascularity in the endometrium as seen by color Doppler

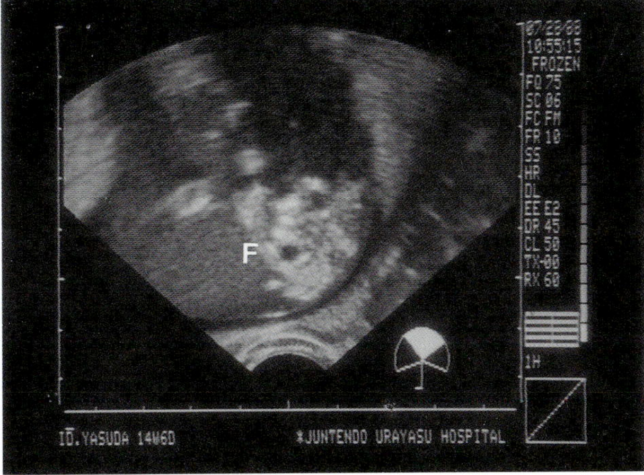

Figure 5.29 The characteristic morphology of the face of an anencephalic fetus at 14 weeks of pregnancy is well depicted in an echogram. F = face

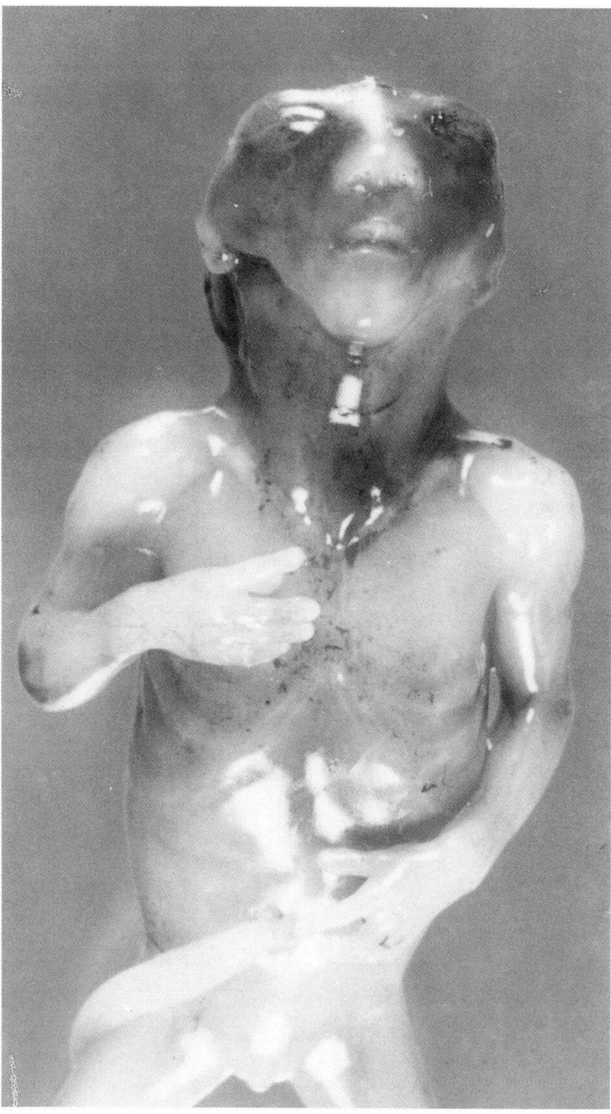

Figure 5.30 The face of the anencephalic fetus shown in Figure 5.29

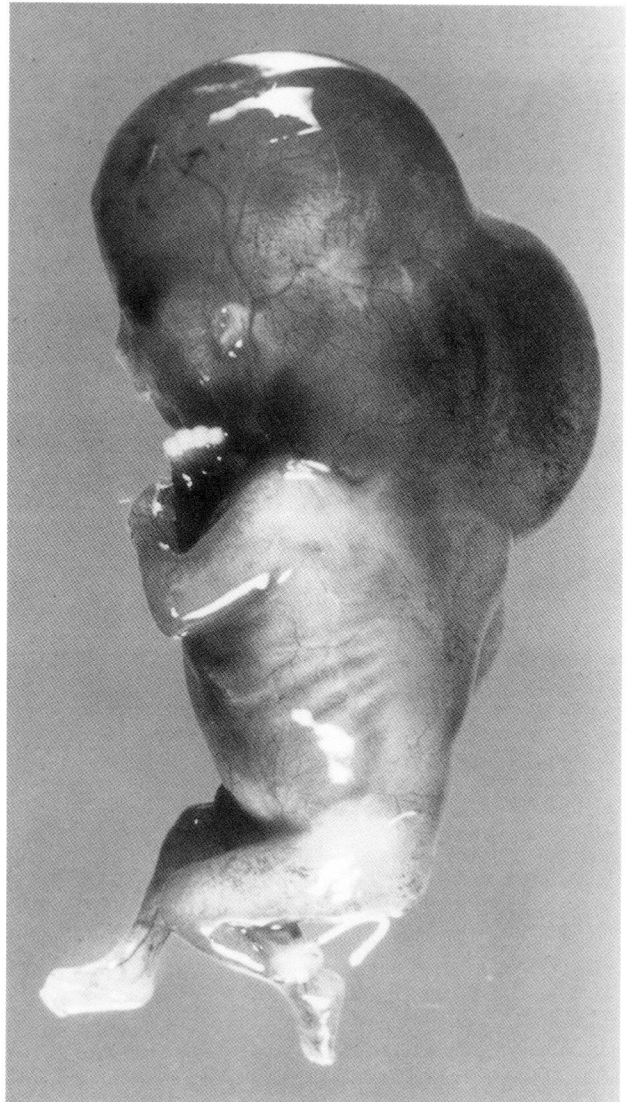

Figure 5.31 Cystic bulge of hind neck

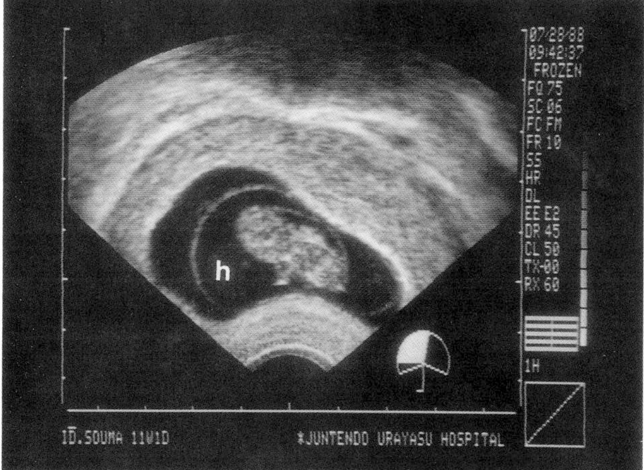

Figure 5.32 The cystic bulge (h) is clearly shown on an echogram

performing non-invasive studies of circulation in pregnancy under physiological circumstances[18].

Transvaginal sonography offers the means to clearly evaluate the embryonal structures. By this extremely sensitive method, it is possible to differentiate the head of the embryo, which makes up almost half the length of an 8-week embryo.

The value of Doppler studies in pregnancy is well known and documented in the literature[21,22]. However, experience in early pregnancy is very limited[23,24]. Ultrasound examination of pelvic circulation by a transvaginal probe combined with pulsed and color Doppler assessment may increase the reliability of ultrasound diagnosis in the early period of pregnancy. Embryonic and fetal vessels can be easily visualized by transvaginal color Doppler, and the pulsed Doppler beam can thus be easily guided to the vessels of interest. If only pulsed Doppler is used, the localization of blood flow is time consuming and a relatively difficult procedure. The guidance of a pulsed Doppler beam by transvaginal color Doppler helps to locate areas of most abundant flow and thus makes examination much faster and more accurate.

The transvaginal color Doppler technique has enabled us to demonstrate, in detail and with accuracy, measurements of blood flow through the vessels of the embryo, uterus and ovaries, which up until now have not been measurable by any known existing diagnostic method in the 5th week of pregnancy. The results obtained by the transvaginal color Doppler sonography method in the group of patients with early pregnancy failure clearly show that the analysis of blood flow in major embryonic vessels can be performed as early as the 6th week of gestation.

The characteristics of blood flow in the aorta, intracranial and umbilical arteries show no diastolic flow during the first trimester.

Further investigation will affirm transvaginal color imaging as a diagnostic tool for the evaluation of normal morphological and functional features of early pregnancy. In the future, transvaginal color flow mapping will become a part of the routine work in the examination of early pregnant women.

REFERENCES

1. D'Addario, V. and Kurjak, A. (1986). Embryologic foundation of the first trimester ultrasonography. In Kurjak, A. (ed.) *Diagnostic Ultrasound in Developing Countries*, p. 59. (Zagreb: Mladost)
2. Moore, K.L. (1982). *The Developing Human*, p. 70. (Philadelphia: Saunders)
3. Hansmann, M., Hackeloer, B.-J. and Staudach, A. (1985). *Ultrasound Diagnosis in Obstetrics and Gynecology: Pregnancy (First Trimester)*, p. 35. (Berlin, Heidelberg, New York and Tokyo: Springer Verlag)
4. Kurjak, A., Žalud, I. and Crvenković, G. (1989). Transvaginal color Doppler in the assessment of maternal and fetal circulation. *International Symposium on Transvaginal Ultrasonography*, Rotterdam, November 2–4, p. 93 (abstr.)
5. Blumenfeld, Z., Rottem, S., Elgali, S. and Timor-Tritsch, I.E. (1988). Transvaginal sonographic assessment of early embryological development. In Timor-Tritsch, I.E. and Rottem, S. (eds.) *Transvaginal Sonography*, p. 87. (New York: Elsevier)
6. Stabile, I., Campbell, S. and Grudzinkas, J.G. (1989). Doppler ultrasound and early pregnancy failure. *Lancet*, **1**, 910
7. Kurjak, A. (1986). *Atlas of Ultrasonography in Obstetrics and Gynecology*, p. 8. (Zagreb: Mladost)
8. Kurjak, A., Wibisono, W. and Jurković, D. (1986). Ultrasound in early pregnancy. In Kurjak, A. (ed.) *Diagnostic Ultrasound in Developing Countries*, p. 65. (Zagreb: Mladost)
9. Van Dongen, L.R.G. and Goudie, E.G. (1980). Fetal movement patterns in the first trimester of pregnancy. *Br. J. Obstet. Gynaecol.*, **87**, 191
10. Alfirevic, Z. and Kurjak, A. (1990). Transvaginal color Doppler ultrasound in normal and abnormal early pregnancy. *J. Perinat. Med.*, **18**, 173
11. Kushnir, O., Izquierdo, L., Vigil, D. and Curet, L.B. (1990). Early transvaginal sonographic diagnosis of gastroschisis. *J. Clin. Ultrasound*, **18**, 194
12. Sauerbrei, E., Cooperberg, P.L. and Poland, B.J. (1980). Ultrasound demonstration of the normal fetal yolk sac. *J. Clin. Ultrasound*, **8**, 217
13. Green, J.J. and Hobbins, J.C. (1988). Abdominal ultrasound examination of the first-trimester fetus. *Am. J. Obstet. Gynecol.*, **159**, 165–75
14. Hata, K., Hata, T., Aoki, S., Hirayama, K., Takamiya, O. and Kitao, M. (1987). Evaluation of female intrapelvic blood vessels and their hemodynamics with real-time two-dimensional Doppler ultrasound. *Nippon-Sanka-Fujinka-Gakkai-Zasshi*, **40**, 67
15. Schulman, H., Ducey, J. and Farmakides, G. (1987). Uterine artery Doppler velocimetry: the significance of divergent systolic/diastolic ratios. *Am. J. Obstet. Gynecol.*, **157**, 539
16. Deutinger, J., Rudelstorfer, R. and Bernaschek, G. (1988). Vaginosonographic velocimetry of both main uterine arteries by visual vessel recognition and pulsed Doppler method during pregnancy. *Am. J. Obstet. Gynecol.*, **159**, 1072
17. Timor-Tritsch, I.E., Rottem, S. and Blumenfeld, Z. (1988). Pathology of the early intrauterine pregnancy. In Timor-Tritsch, I.E. and Rottem, S. (eds.) *Transvaginal Sonography*, p. 109. (New York: Elsevier)
18. Kurjak, A., Alfirevic, Z. and Miljan, M. (1988). Conventional and color Doppler in the assessment of fetal and maternal circulation. *Ultrasound Med. Biol.*, **14**, 337
19. Kurjak, A., Žalud, I. and Crvenković, G. (1990). Transvaginal color Doppler in normal and abnormal early pregnancy. Presented at the *7th Congress of the European Federation of Societies for Ultrasound in Medicine and Biology*, Jerusalem, May 5–10, Abstr. p. 71
20. Enk, L., Wikland, M., Hammarberg, K. and Lindblom, B. (1990). The value of endovaginal sonography and urinary human chorionic gonadotropin tests for differentiation between intrauterine and ectopic pregnancy. *J. Clin. Ultrasound*, **18**, 73
21. Cohen-Overbeek, T., Pearce, J.M. and Campbell, S. (1985). The antenatal assessment of utero-placental and fetoplacental blood flow using Doppler ultrasound. *Ultrasound Med. Biol.*, **11**, 329
22. D'Agincourt, L. (1988). Doppler opened new paths into fetal evaluation. *Diagnostic Imaging*, **10**, 110
23. McSweeney, M.B., Warren , P.S., Baber, R.J., Kossoff, G. and Gill, R.W. (1987). Transvaginal Doppler assessment of pelvic vascularity in early pregnancy: a collaborative study. *Presented at the Australian Society for Ultrasound in Medicine, 17th Annual Scientific Meeting*, Canberra, p. 38
24. Schaaps, J.P. and Soyeur, D. (1989). Pulsed Doppler on a vaginal probe. Necessity, convenience, or luxury? *J. Ultrasound Med.*, **8**, 315

6 Assessment of Placental Development and Function

E. Jauniaux, D. Jurković, A. Kurjak and J. Hustin

INTRODUCTION

Most of the present knowledge of the placental circulation is derived from perfusion of human placenta *in vitro* or from animal experiments, in particular in the pregnant Rhesus monkey[1]. Before the advent of ultrasound techniques, data close to the *in vivo* situation in humans could only be obtained when a hysterectomy was performed during pregnancy.

The use of color Doppler ultrasound techniques combined with transvaginal sonography offers a novel approach for the investigation of human uteroplacental and fetal circulations[2].

Pregnancy is associated with major hemodynamic changes in the maternal circulation associated with continuous growth and development of the placental circulation. Knowledge of the normal anatomy and physiology of the placenta is a prerequisite for the development of pathophysiological hypotheses which will lead to a better understanding of pregnancy disorders affecting the placental circulations.

THE ANATOMY AND CIRCULATIONS OF THE PLACENTA

Development of the placenta and the fetal circulation

The placenta has two components: a fetal portion developed from the embryonic chorion and a maternal portion formed by the pregnant endometrium or decidua. Formation of the placenta begins when implantation begins. As soon as the blastocyst attaches to the endometrium, trophoblastic infiltration begins[1]. The entire blastocyst sinks progressively into the endometrial stroma and the migrating trophoblast encounters venous channels of increasing size, then superficial arterioles and eventually, during the second week, the endometrial spiral arteries[1, 3–5].

The formation of primary chorionic villi begins between 13 and 15 days after ovulation, corresponding to stage 6 of embryonic development[5]. Simultaneously, blood vessels start in the extra-embryonic mesoderm of the yolk sac, the connecting stalk and the chorion[5]. By 18–21 days (stages 8 and 9), the villi have become branched and the mesenchymal cells within the villous mesoblastic core have differentiated into blood capillaries, forming an arteriocapillary venous network[5]. These vessels soon become connected with the primitive heart tubes and the vascular plexus of the yolk sac by the developing vessels of the connecting stalk[5]. At the beginning of embryonic developmental stage 10 (21 days or the beginning of the 4th week postovulation), the primitive heart begins to beat and an ebb-and-flow circulation begins between the embryo and the placenta[5].

Up to the 8th week (around 10 weeks after the date of the last menstruation) which corresponds to the last week of the embryonic period (stage 23), villi cover the entire surface of the chorionic sac[3–5]. As the sac grows, the villi associated with the decidua capsularis, surrounding the amniotic sac, are compressed and begin to degenerate, forming an avascular shell known as the chorion laeve or smooth chorion. By the end of the first trimester, the enlarging gestational sac fills the entire uterine cavity and fusion of the decidua capsularis and the decidua parietalis obliterates completely the uterine cavity[3–5].

The gross anatomy of the human placenta is basically the anatomy of its fetal circulation with its connective tissue supports and trophoblastic covering. The maternal tissues contribute little to the architecture of the definitive placenta. At the end of the first trimester, the placenta can be clearly separated into three basic structures: the chorionic or fetal plate, the placental villous tissue or substance, and the basal or maternal plate[3, 4].

The mature placental mass is divided into 20–40 cotyledons or functional units, determined by the branching pattern of the villous tree. These units include the primary villous trunk with its derivatives forming 180–320 subunits or lobules[1, 3, 4]. The functional units must be differentiated from the placental lobes, familiar to clinicians as distinct lobulation on the maternal surface of the placenta after delivery. These lobes are formed by septa arising from the basal plate towards the chorionic plate and have no physiological significance[3, 4].

Development of the maternal placental circulation

The uterine arteries originate from the internal iliac arteries. The main uterine arteries give off branches which extend inward for about a third of the thickness of the myometrium without significant branching and then subdivide into an arcuate wreath encircling the uterus [1]. From this network arise smaller branches, the radial arteries, directed towards the uterine lumen. These arteries become the endometrial spiral arteries as they pass the myometrial–endometrial junction [1,6]. The human uterine circulation is very rich in arterial anastomosis. Branches of the uterine arteries anastomose with branches of the ovarian and vaginal arteries to establish a vascular arcade perfusing the internal genital organs [1,6]. Vascular connections are also found between the main uterine arteries and their largest branches or joining the uterine circulation with the systemic circulation [6].

Human placentation is hemochorial and mainly characterized by diffuse infiltration of the placental bed by extravillous trophoblast [7,8]. The migratory trophoblast penetrates both decidua and the spiral arteries [7,8]. At about 10–12 weeks of gestation (menstrual weeks), this trophoblastic wave reaches the deciduo–myometrial junction [7,8]. During the 2 first months of the second trimester, the migratory trophoblast spreads down into the myometrial segments of the spiral arteries until their origins from the radial arteries. These physiological changes play a key role in converting the spiral arteries into distended, low-resistance uteroplacental arteries [1,7,8].

The current hypothesis concerning early human placentation is that, within a few days after implantation, spiral arterioles are trapped by the migratory trophoblast and a true maternal blood flow starts [1,3,5]. This hypothesis is not supported by recent perfusion experiments of hysterectomy specimens with pregnancy *in situ* at various gestational ages [9,10]. Examination of slice radiographs of specimens collected at 7, 8 and 9 weeks of gestation shows no contrast medium in the intervillous space (Figure 6.1), while at 13 weeks the placenta is rapidly filled with the contrast medium. Furthermore, pathological investigations of these hysterectomy specimens demonstrate complete occlusion of the vascular lumen of the tip of the uteroplacental arteries by trophoblastic cells in early pregnancies (Figure 6.2). Reconstruction from serial sections of the course of spiral arteries in these cases is also suggestive of the absence of blood flow in the intervillous space before 12 weeks of gestation (Figure 6.3). At 13 weeks of gestation, on the contrary, the uteroplacental arteries are free of trophoblastic plugs and contrast medium is found in the intervillous space, encircling the chorionic villi. These data were confirmed *in vivo* by means of hysteroscopy, examination of chorionic villous sampling material and transvaginal sonography [9,11]. From these results, it can be stated that the early placenta before 12 weeks of gestation is bathed by a fluid, possibly composed of maternal plasma and uterine gland secretions, and not by continuous maternal blood flow [9–12].

In utero aspects of the placental developmental anatomy

Transabdominal ultrasonography (TAS)

Visualization and localization of the placenta by TAS has become rapidly superior to all other imaging techniques such as radiographic placentography or scintigraphy and is now an essential part of routine prenatal examinations.

The gestational sac can be observed by means of abdominal gray-scale imaging as early as 42 days from the last menstrual period (embryonic stage 13), when the level of human chorionic gonadotropin (β-hCG) in the maternal serum is greater than 6500 mIU/ml [13]. However, the definitive diagnosis of a normal intrauterine pregnancy is postponed until the fetal pole is seen around 6–7 menstrual weeks [13,14].

Sonographic evidence of placental textural changes *in vivo* has been reported during the second half of pregnancy [15–18]. A sonographic classification system for grading placentas *in utero* according to maturational changes was developed by Grannum and associates [16]. The placentas were graded from 0 to III (Figure 6.4), on the basis of compound B-scan changes in the placental structures, and these results were correlated with fetal pulmonary maturity evaluated by amniotic fluid lecithin–sphingomyelin (L/S) ratios.

Some of these authors have attempted to reproduce *in vitro* the sonographic morphological features observed *in vivo* [17,18]. In these studies, however, the placentas were not always fixed immediately after delivery and some changes observed *in vivo* were not found *in vitro* and new features could have been created during placental collapse. More satisfactory results were obtained using a perfusion system *in vitro*, to avoid placental collapse [19,20]. Sonolucent areas or echo-free spaces frequently observed in the centre of the cotyledon or under the fetal plate of the placenta during the second half of pregnancy (Figure 6.5) have been correlated with avillous zones [19–22], and the peripheral echoes with septal calcification and fibrin deposition [17,18].

The overall morphology of placental vasculature, as demonstrated with color Doppler imaging,

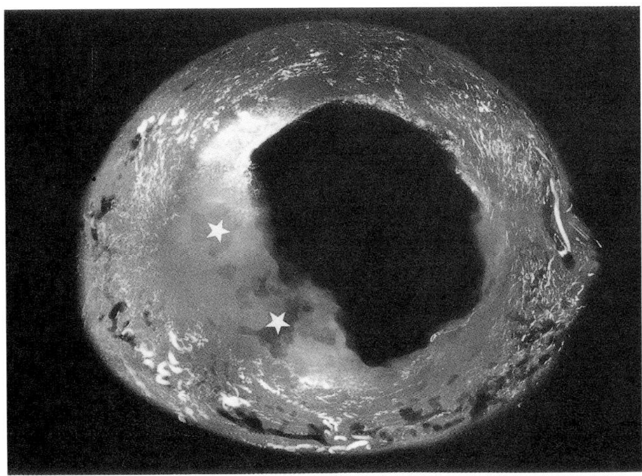

Figure 6.1 X-ray of a 1 cm thick slide through the whole uterus from a pregnancy at 8 weeks' gestation. No contrast medium is found in the placental area (*)

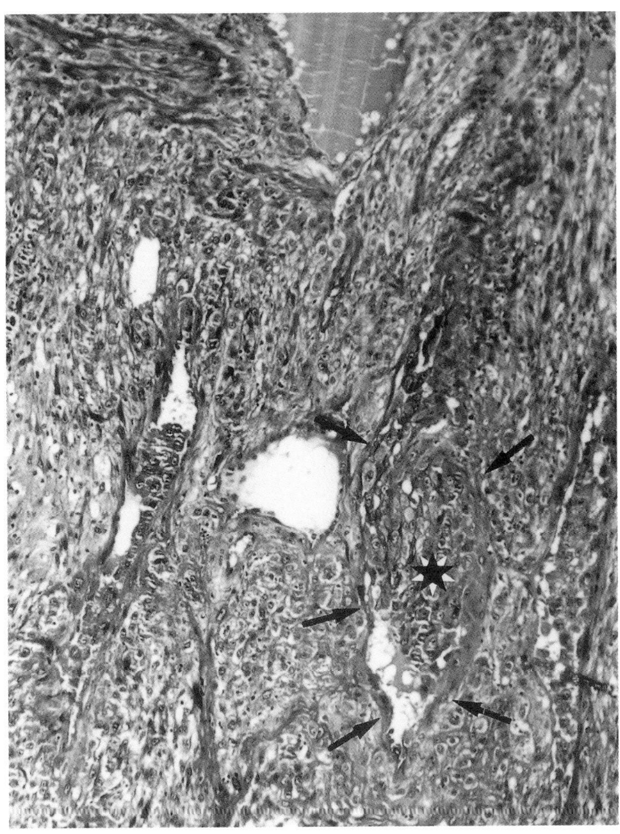

Figure 6.2 Microscopic section of the placental implantation site at 8 weeks' gestation demonstrating a uteroplacental artery (arrows) occluded by a trophoblastic plug (*) in continuity with the trophoblastic shell (H & E x 40)

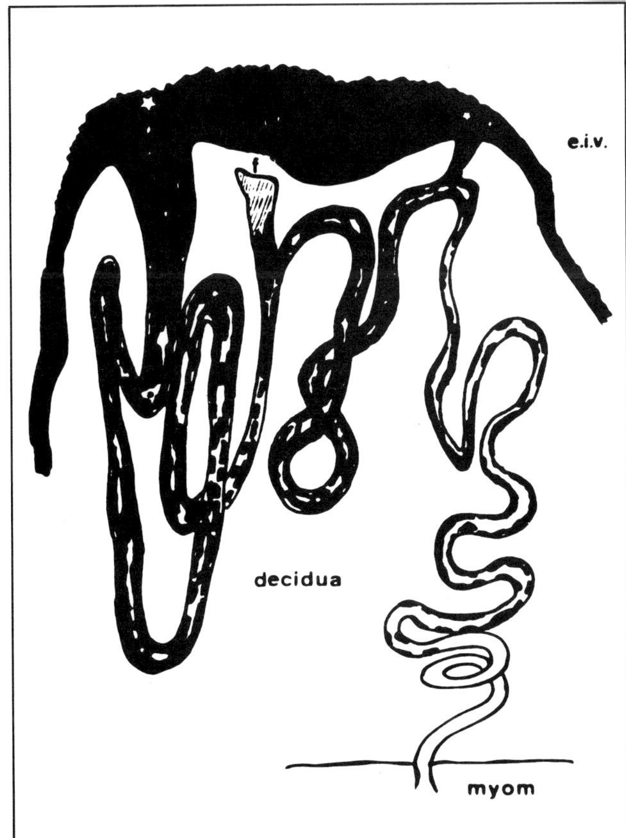

Figure 6.3 Reconstruction of the course of a spiral artery based on serial sections of a hysterectomy specimen containing a 9 weeks' pregnancy. Average length irrespective of numerous coils varies between 0.5 and 0.7 mm. No contrast medium is observed in the intervillous space (e. i. v.)

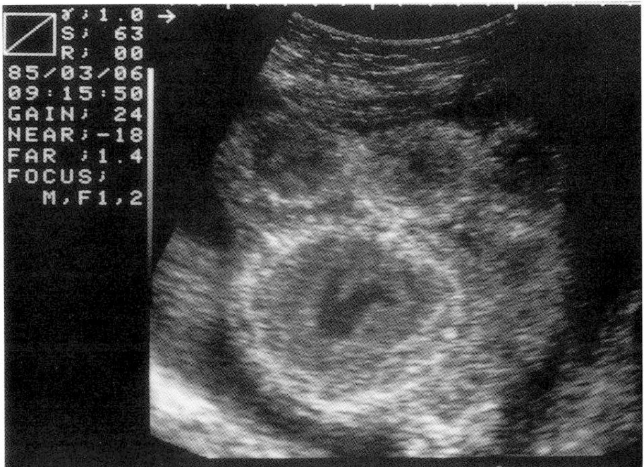

Figure 6.4 Premature grade III at 32 weeks in a pregnancy complicated by chronic hypertension

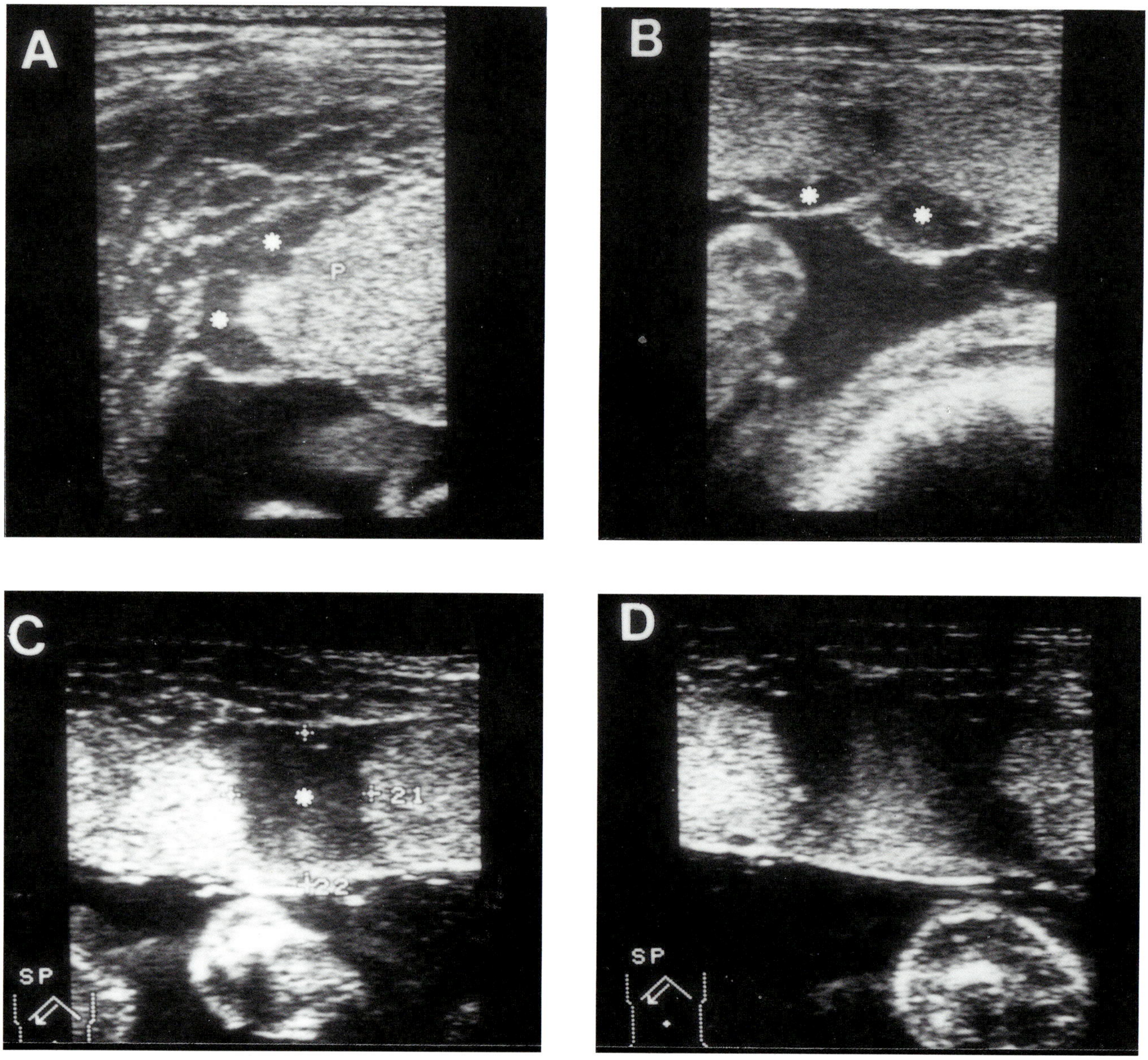

Figure 6.5 Sonograms of placental sonolucent areas (*). **A** Margin between the placenta and the uterine draining venous plexus; **B** under the umbilical cord insertion; **C** and **D** transverse and longitudinal views of a large sonolucent space containing turbulent blood flow

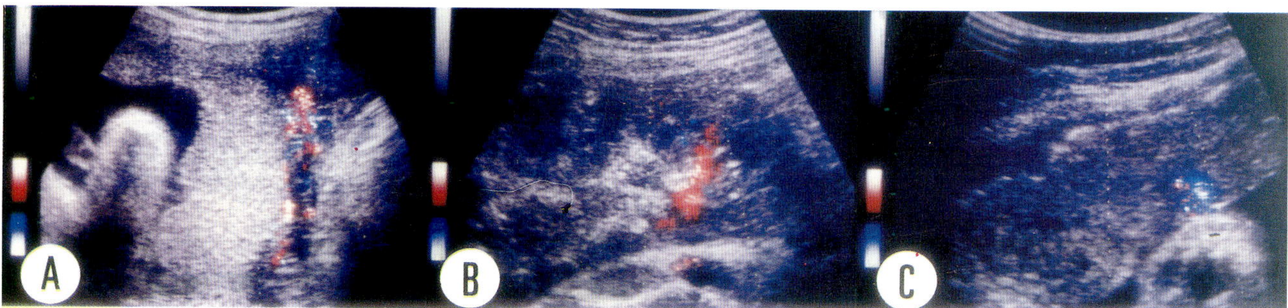

Figure 6.6 Color flow imaging of the placental bed at 18 weeks' gestation (**A**) and large intervillous spaces at term (**B** and **C**)

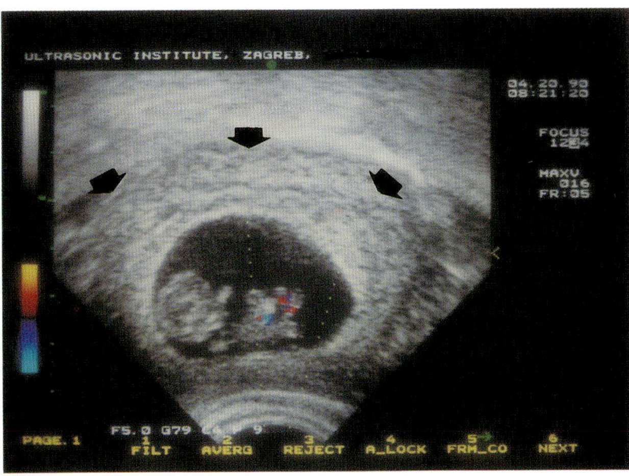

Figure 6.7 Transvaginal color flow imaging at 10 weeks' gestation showing the placenta (arrows) and fetal flow signals

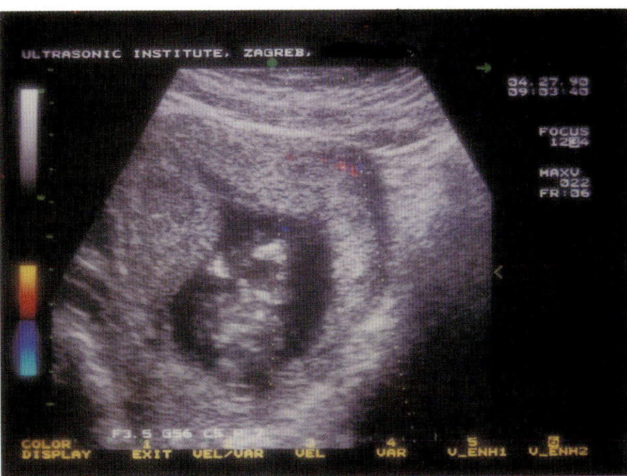

Figure 6.8 Abdominal color flow imaging at 13 weeks' gestation showing clearly the definitive placenta

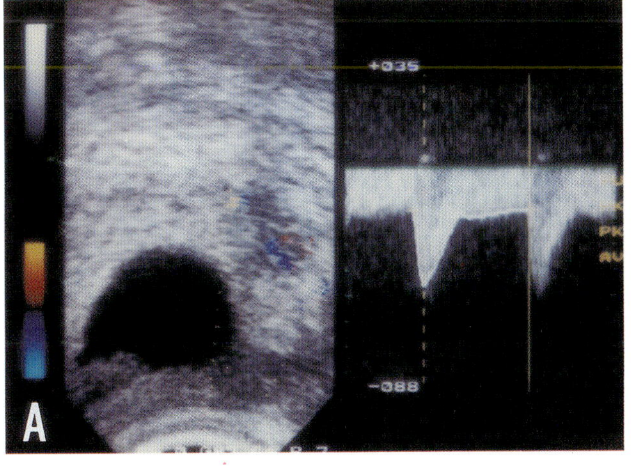

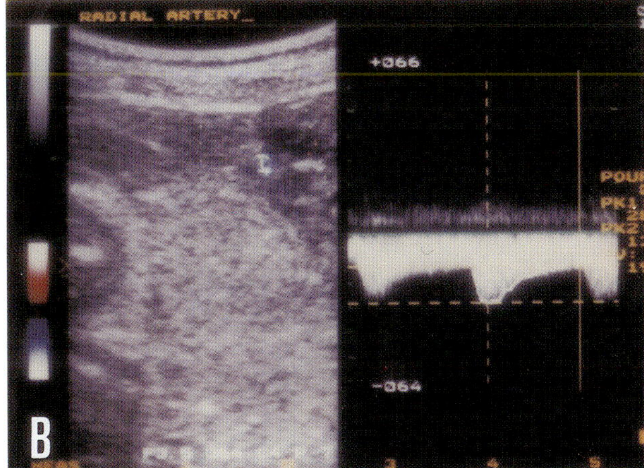

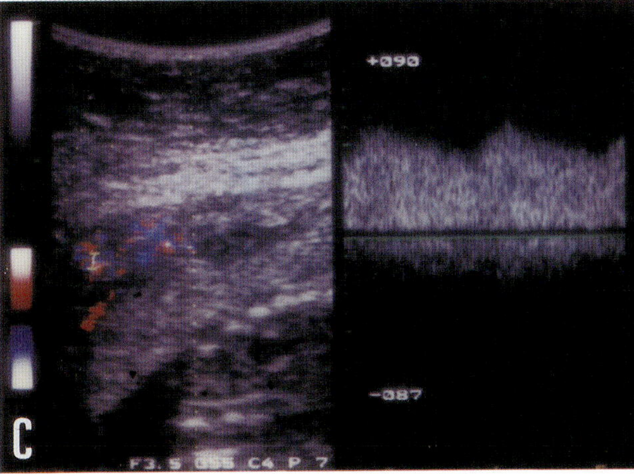

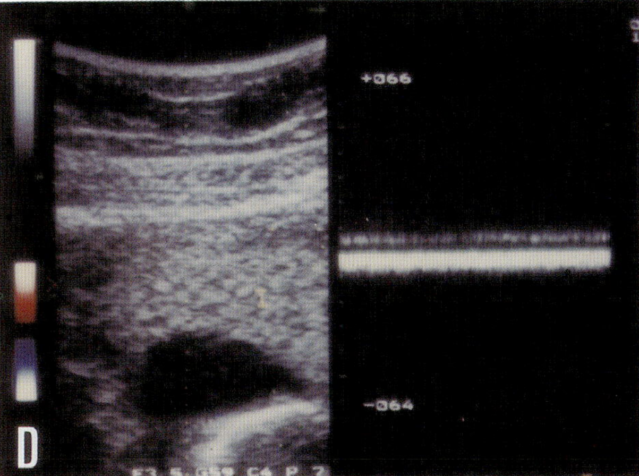

Figure 6.9 Transvaginal and abdominal color Doppler ultrasound and spectral analysis, at 18 weeks' gestation, of the maternoplacental circulation waveforms obtained from arcuate (**A**), radial (**B**), and uteroplacental (**C**) arteries and from maternal intervillous circulation (**D**)

correlates well with the classical anatomic findings. Sonolucent areas identified by real-time imaging contain turbulent flow on color imaging with a venous pattern on spectral analysis (Figure 6.6).

Transvaginal ultrasonography (TVS)

The recent rediscovery of the vaginal route for sonography in reproductive medicine has opened a new field of investigation for sonographers oriented to fetal and placental development. The first demonstration of an intrauterine pregnancy by means of TVS was reported by Kratochwill in 1967 [23]. With the major improvements in ultrasonographic resolution in the early 1980s, TVS became a method used world-wide for detailed investigations of early pregnancy.

The gestational sac is detectable by means of TVS around 4–5 menstrual weeks as a small echogenic ring [9, 11, 12, 24–28]. Recent studies have correlated the TVS features with levels of β-hCG in the maternal serum [24–26]. Normal intrauterine pregnancies can be imaged by TVS as early as 32 days after the last menstruation (embryonic stage 8), corresponding to a gestational sac > 3 mm and hCG levels > 750 mIU/ml (International Reference Preparation). Therefore, TVS appears also to be a very promising non-invasive method for the early and accurate diagnosis of extrauterine pregnancy [27, 28].

After 8 weeks of gestation the placenta appears as a thickening of a portion of the gestational sac (Figure 6.7). The different structures forming the definitive placenta can be clearly discerned *in vivo*, at the end of the first trimester (Figure 6.8). Conventional TVS allows a clear definition of anatomic structures of 1 mm width in the explored field [9, 11, 12]. Small moving echoes synchronous with the maternal pulse can be identified in uterine vessels, on real-time imaging, at approximately 1 cm apart from the basal plate of the placenta, but no continuous blood flow could be detected in the trophoblast area before 12 weeks of gestation [9, 11].

Color flow imaging is revolutionizing our ability to map out the fetal and maternal components of placental circulations. Identification of the various branches of the main uterine arteries and the umbilical cord circulation can be easily made, from 5 and 8 weeks of gestation, respectively (Figures 6.9 and 6.10). The two-dimensional display allows accurate placement of the Doppler gate for spectral analysis. Intraplacental waveforms with fetal or maternal characteristics can be identified and clearly differentiated from 15 weeks of gestation (Figures 6.9D and 6.10D).

PLACENTAL CIRCULATORY PHYSIOLOGY

Continuous circulation of maternal blood through the intervillous chamber is a dynamic process that requires a continuous adaptation of the individual cotyledon to the blood flow offered to it by the corresponding uteroplacental artery. Compromises in the placental circulatory homeostasis are associated with significant fetal complications resulting from disruption of an adequate and continuous supply of oxygen and nutrients.

Hemodynamic changes in the maternal circulation associated with pregnancy

During the first half of normal pregnancies the maternal blood volume and cardiac output increase progressively by an average of 40% above non-pregnant values [29, 30]. Wide individual variations have been observed for these changes and they are probably not directly related, as the blood volume plateaus after the 30th week of gestation while the cardiac output reaches a peak by 20–24 weeks and then decreases towards term [29, 30].

The progressive elevation in uterine blood flow is also out of phase with the augmentation of maternal cardiac output. Uterine blood flow rises to only 100 ml/min at the end of the first trimester, reaches 200 ml/min by the 28th week and is estimated to be around 500 ml/min at the end of pregnancy [11, 29]. Therefore, until midterm, the rise in uterine blood flow accounts for only a portion of the increment in cardiac output.

As discussed previously, during the first trimester of pregnancy, the growing embryo and its placenta appear to be separated from the maternal circulation [9, 12]. After 12 weeks, the trophoblastic plugs in the spiral arteries, still remaining from the first extravillous trophoblastic wave, no longer obliterate the uteroplacental arteries and a real intervillous circulation is then established. The endovascular migratory trophoblast now invades the deep segments of the spiral arteries, allowing the progressive dilatation of these vessels. Color Doppler ultrasound shows no change in the characteristics of the waveforms obtained from the uteroplacental arteries at different gestational ages (Figures 6.11 and 6.12). A lower resistance and turbulences (arteriovenous like) are always shown compared with the main uterine arteries (Figure 6.13) or the waveforms of their larger branches (Figures 6.9A and 6.9B). In all segments of the maternal placental circulation, the resistance indexes decline with advancing gestational age and towards the placenta.

A fall in uterine vascular resistance is not only due to erosion of the spiral vessels by the invasive trophoblast, as it occurs also in animal species, in which no erosion of the maternal vessels occurs[29]. Variations in circulating steroid and protein hormones also play an important role in maternal hemodynamic changes encountered during normal pregnancy in human.

Hemodynamic changes in the fetal circulation

During prenatal life, the deoxygenated blood leaves the fetus and reaches the placenta via the umbilical arteries. These arteries divide into smaller vessels (chorionic vessels) running freely on the fetal plate of the placenta, before entering inside the placental mass and subsequently in the main villous trunks. From this network arise smaller branches, forming an extensive arteriovenous system within the terminal villi, bringing fetal blood in very close contact with the maternal intervillous circulation[1,3–5]. The oxygenated blood is carried to the fetus via the umbilical vein. The fetal blood is poorly oxygenated compared to the adult values and lower pO_2 and pH values have been demonstrated *in utero* in the umbilical vein compared to the intervillous space[31]. Furthermore, the fall of pO_2 in the umbilical vein with progressive gestation is parallel with the fall in pO_2 in the intervillous space, suggesting that the placenta is a major site of oxygen consumption[31].

As for the placental maternal circulation, Doppler signals can be obtained from the different segments of the fetal circulation (Figure 6.10), and spectral analysis also shows a decline in resistance indexes with advancing gestation and towards the intraplacental arterioles. However, the waveforms obtained from the fetal circulation have different characteristics before and after 14 menstrual weeks, due to the physiological absence of an end-diastolic flow before 14 weeks (Figures 6.14 and 6.15).

The morphology of unfixed placental tissue deteriorates rapidly after delivery[32]. Within a few minutes of total ischemia, alterations of the rough endoplasmic reticulum and the mitochondria are demonstrated in the syncytiotrophoblastic layer (Figure 6.16). Therefore, perfusion fixation should be used more often for pathophysiological studies of placental function related to anatomy[33].

USE OF COLOR DOPPLER IMAGING TO STUDY THE PATHOPHYSIOLOGY OF PLACENTAL AND CORD ABNORMALITIES

TVS will be increasingly used for the detection of fetal and placental abnormalities. As an example, recent comparative series have shown that TVS is equally as safe and diagnostically superior to TAS for the prenatal diagnosis of placenta previa[34,35]. However, the diagnostic value of TVS, in particular during the first trimester, should not be overestimated, due to temporary anatomic variations associated with normal embryonic development. Furthermore, the majority of placental vascular lesions develop during the second half of pregnancy[20–22]. Fetal position and placental localization are practical criteria for the use of TVS or TAS.

Because of its ability to provide spatial resolution to Doppler study, color imaging has the potential to facilitate the application of Doppler to the fetal and placental examination[36,37]. It can provide a technique for identifying the location and type of blood flow waveforms in placental and cord abnormalities detected by imaging.

Placental blood flows in complicated pregnancies

Absence or incomplete trophoblastic infiltration of the spiral arteries, together with lesions of acute atherosis, were described in women presenting with pre-eclampsia and/or fetal growth retardation[8]. Immunohistochemical studies have demonstrated the deposition of IgG, IgM and complement in these lesions[38,39]. Acute atherosis of the uteroplacental arteries is often associated with placental chronic villitis and these lesions are similar to the vascular lesions described in renal, cardiac and hepatic homograft rejection[40]. This process could be a morphological expression of a maternal immune attack against placental tissue.

Several authors have reported placental lesions associated with pregnancies complicated by abnormal umbilical artery Doppler indices. A significant relationship has been found between an increased pulsatility index and a decrease in the number of small arterioles of tertiary stem villi, suggesting an obliterative process[41–46]. Major placental vascular lesions were also reported in these cases[45,46]. Accurate identification of the waveforms of the placental circulations (Figure 6.17) appears promising in understanding the pathophysiology of placental lesions associated with abnormal Doppler signals and, in particular, for the study of placental infarcts and thrombosis.

Placental tumors

Abdominal or transvaginal high resolution color Doppler ultrasound has recently made it possible to identify the waveforms obtained from intrauterine masses such as uterine myomas[2]. Knowledge of the characteristics of the expected fetal or maternal

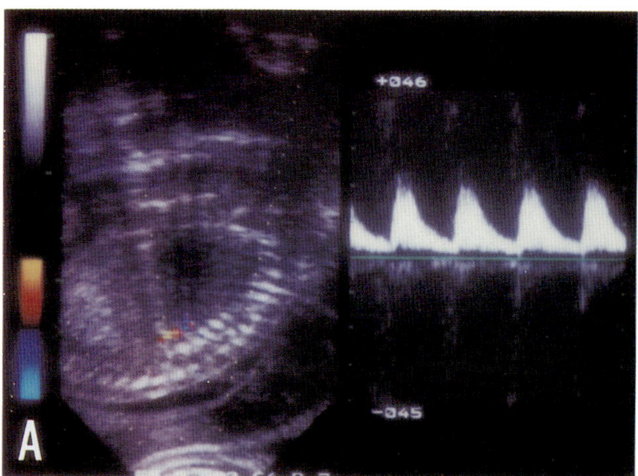

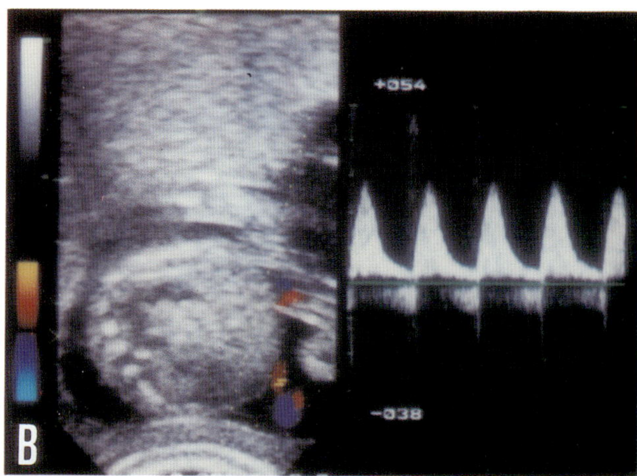

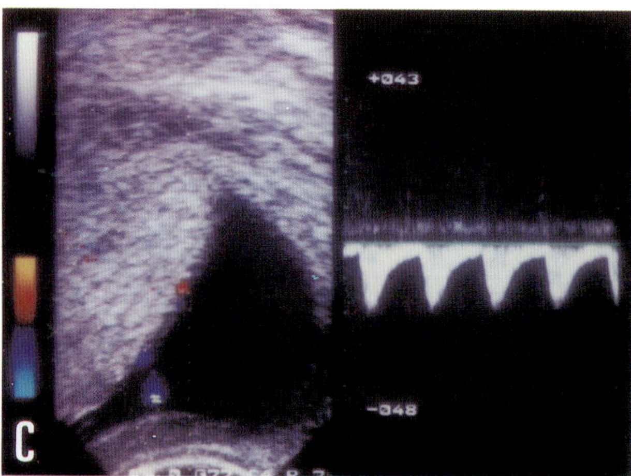

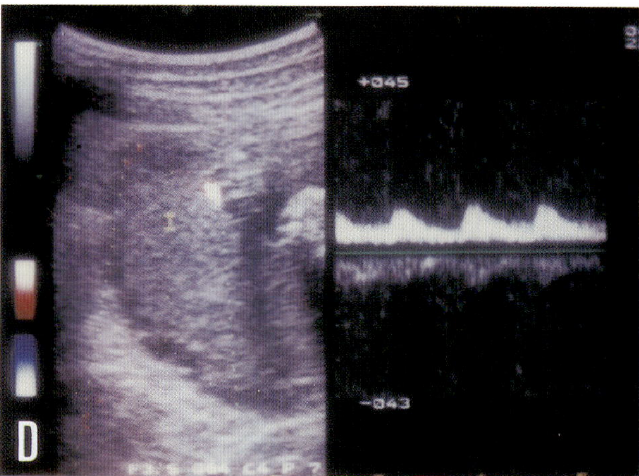

Figure 6.10 Transvaginal and abdominal color Doppler ultrasound and flow velocity waveforms obtained, at 16 weeks' gestation, from the fetal aorta (**A**), the umbilical artery at its fetal (**B**) and placental (**C**) insertion and from an intraplacental arteriole (**D**)

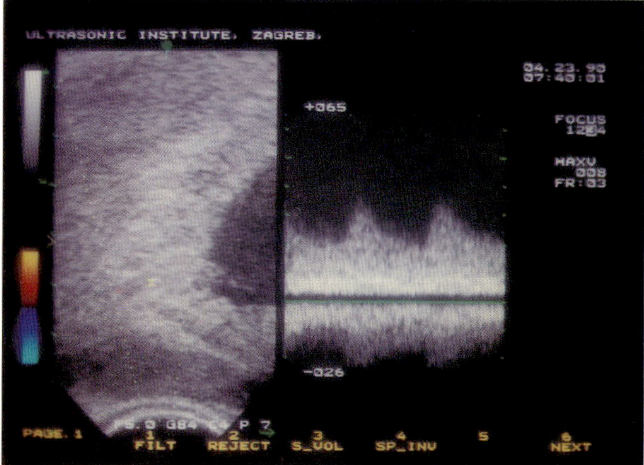

Figure 6.11 Transvaginal color Doppler ultrasound and flow velocity waveforms obtained from the uteroplacental artery near the basal placental plate at 8 weeks' gestation showing a characteristic (turbulent) pattern

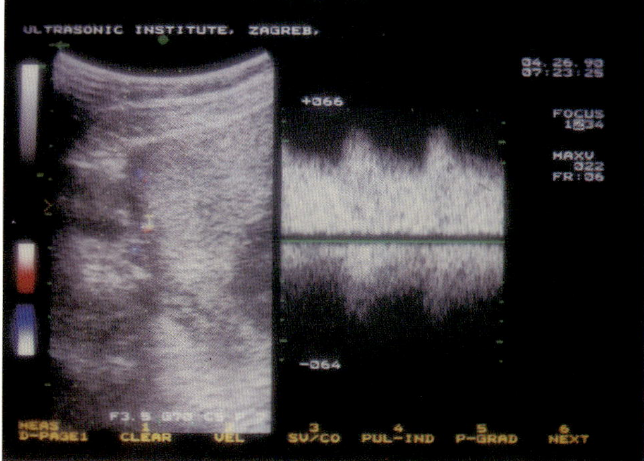

Figure 6.12 Abdominal color Doppler ultrasound and flow velocity waveforms obtained from the uteroplacental artery at 14 weeks' gestation. Compare with Figure 6.11

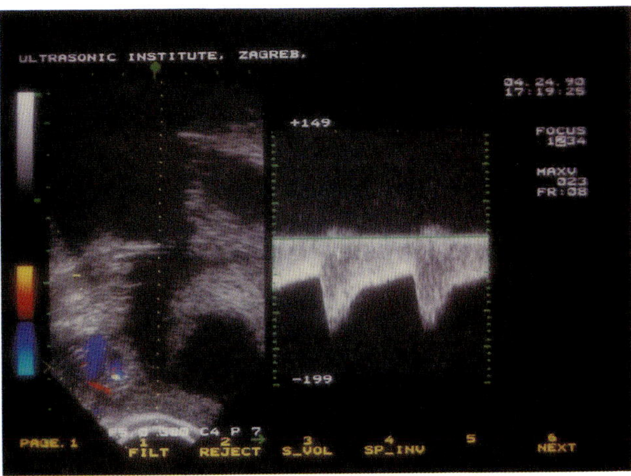

Figure 6.13 Transvaginal color Doppler ultrasound and flow velocity waveforms obtained from the main uterine artery. Compare with Figures 6.11 and 6.12

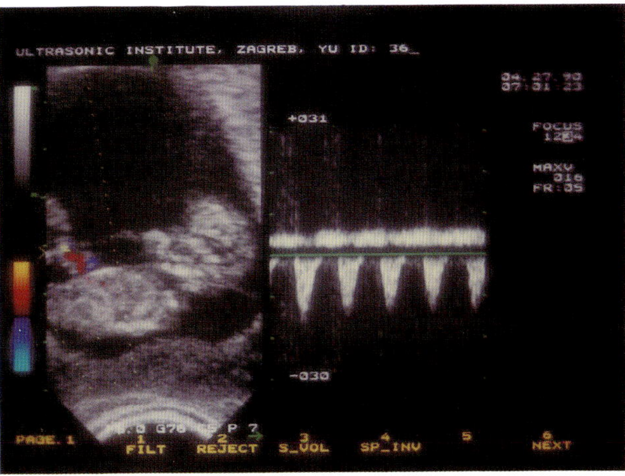

Figure 6.14 Transvaginal color Doppler ultrasound and flow velocity waveforms obtained from the umbilical cord circulation at 11 weeks' gestation demonstrating regular waveform with no end-diastolic flow

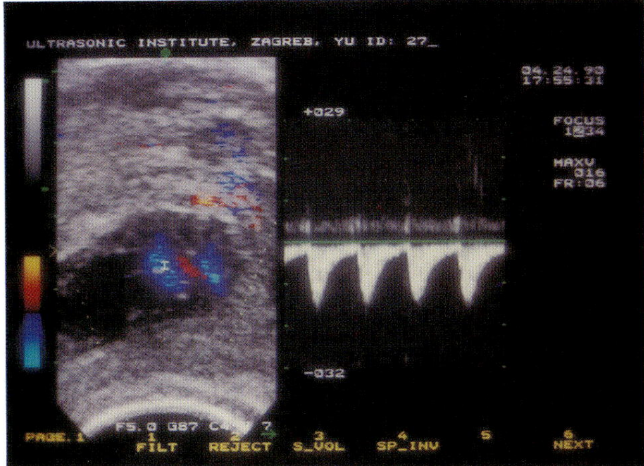

Figure 6.15 Transvaginal color Doppler ultrasound and flow velocity waveforms obtained from the umbilical cord circulation at 15 weeks' gestation. Note the presence of end-diastolic flow. Compare with Figure 6.14

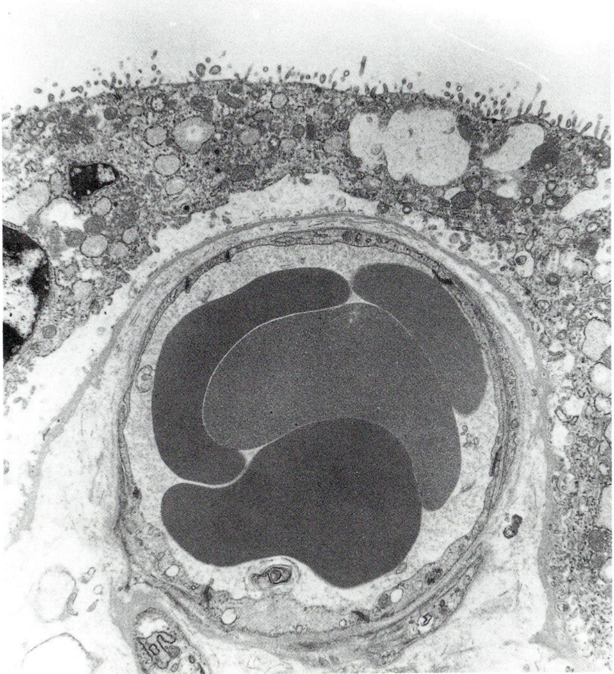

Figure 6.16 Ultrastructural view of a chorionic villus from a normal pregnancy at term fixed immediately after vaginal delivery showing trophoblastic alterations secondary to ischemia

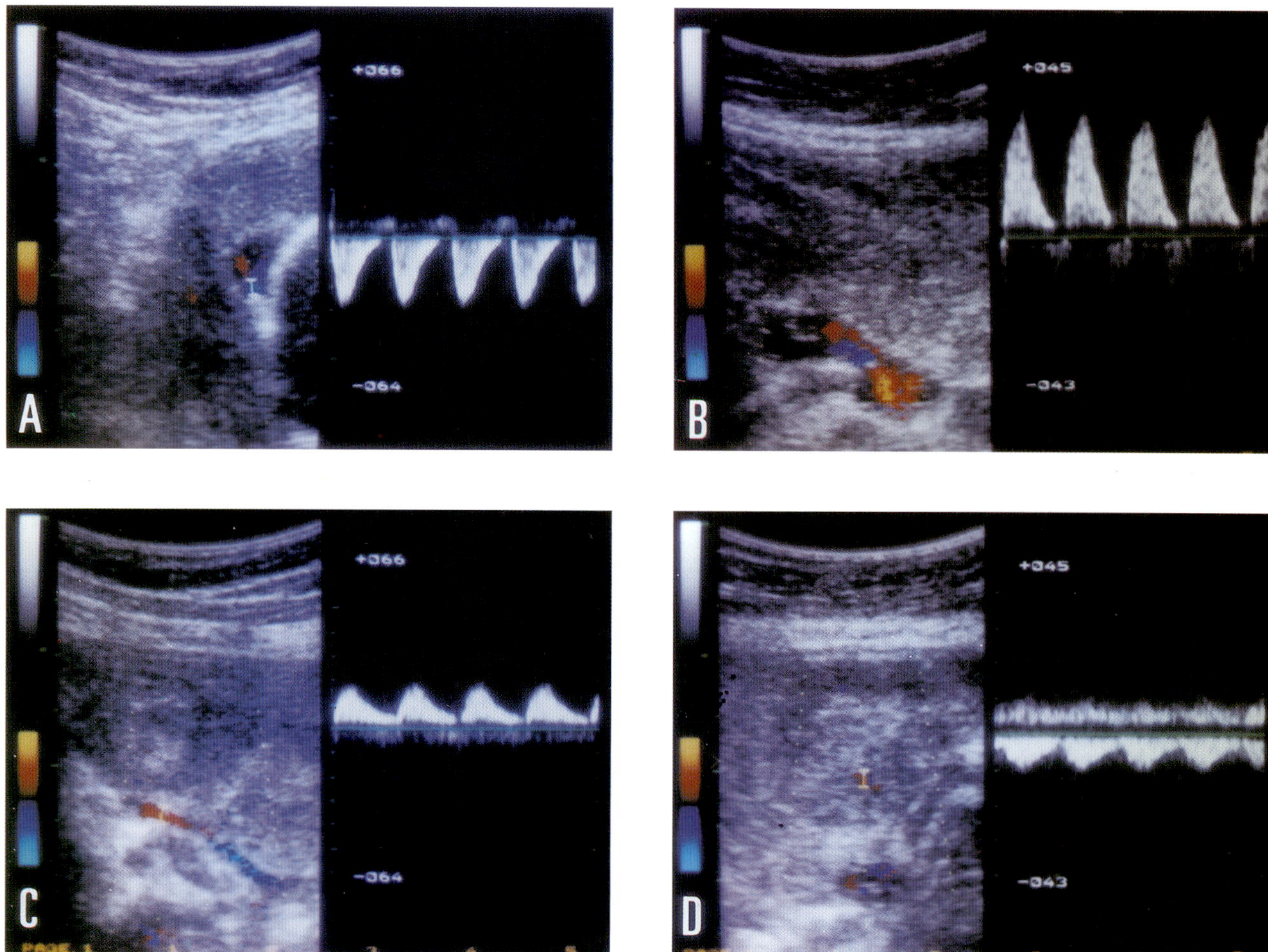

Figure 6.17 Abdominal color Doppler ultrasound and flow velocity waveforms obtained from the fetoplacental circulation of a pregnancy at 26 weeks of gestation, complicated by poor fetal growth, showing no end-diastolic flow in the umbilical artery (**A** and **B**). An end-diastolic flow was present for most of the waveforms obtained at the level of the fetal plate (**C**) and inside the placenta (**D**), probably due to a fall in resistance towards the intraplacental arterioles

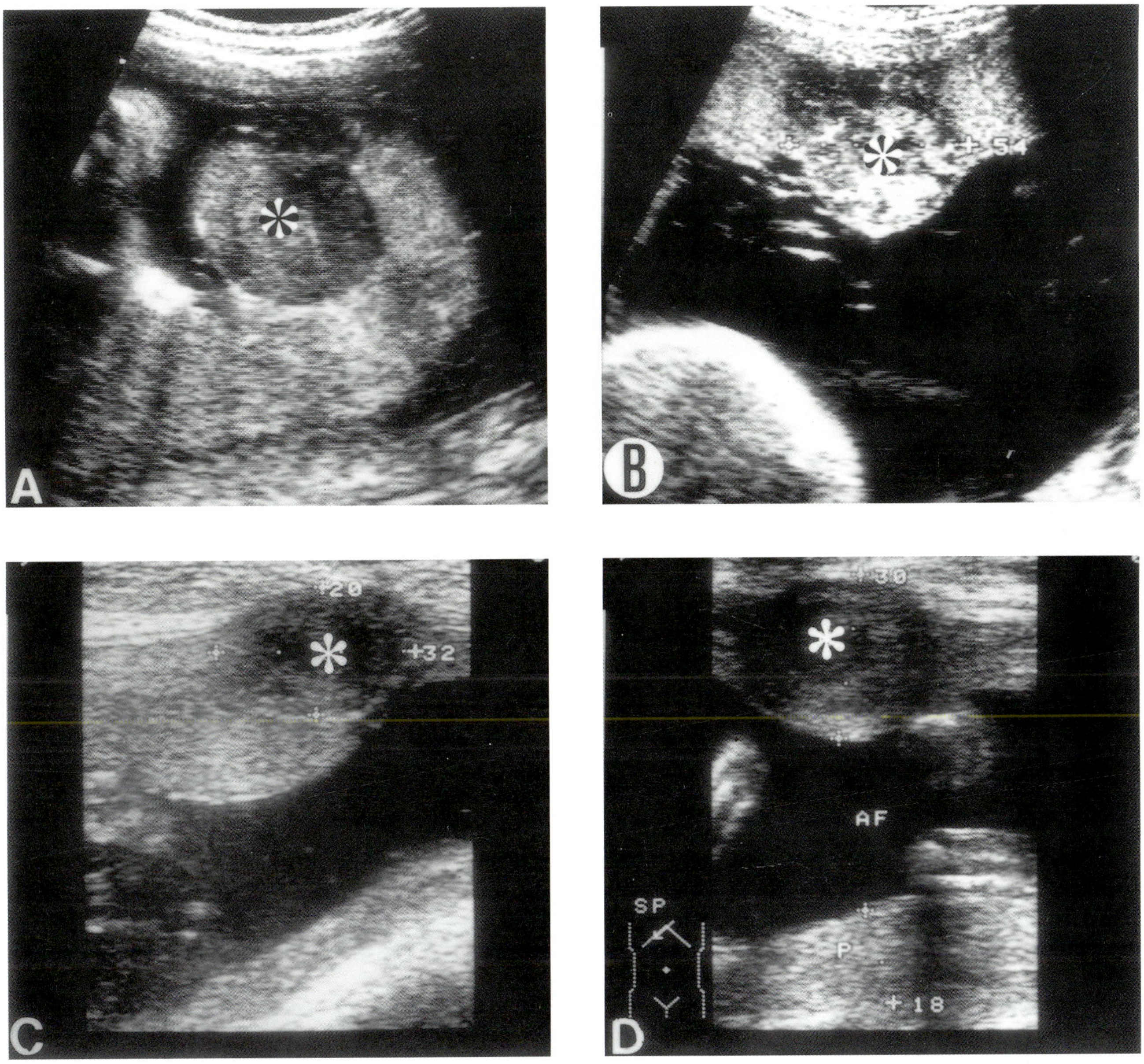

Figure 6.18 Composite compound sonograms of two placental chorioangiomas (**A** and **B**) and two uterine leiomyomas (**C** and **D**). In the case presented in **C**, blood flow waveform analyses were necessary to exclude a marginal chorioangioma of the placenta

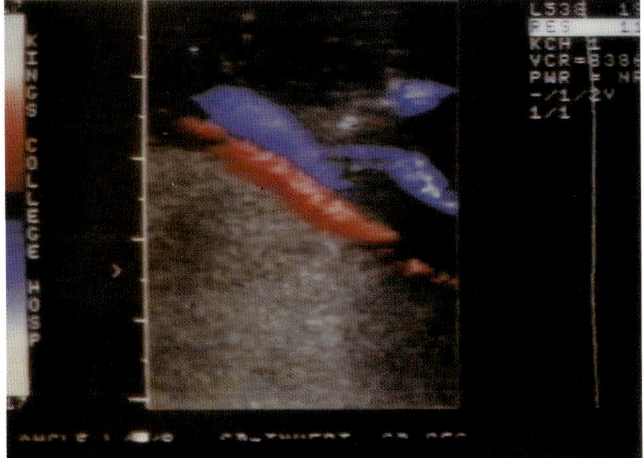

Figure 6.19 Color flow image of a single umbilical artery cord at 18 weeks, associated with multiple fetal malformations and oligohydramnios. (Reproduced with permission of the publisher from Jauniaux *et al.* (1989). *Am. J. Obstet. Gynecol.*, **161**, 1195)

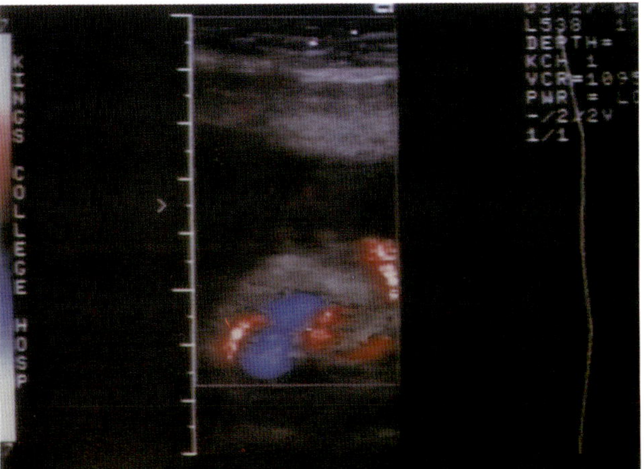

Figure 6.20 Color flow image of the cord tumor at 20 weeks' gestation showing an abnormal vascular pattern at the placental insertion, suggestive of a cord angiomyxoma

waveforms is important for the differential diagnosis of placental tumors (Figure 6.18) from other intrauterine abnormalities. For instance, in cases of undetermined intrauterine tumor, demonstration of the fetal waveforms strongly suggests a chorioangioma, while maternal waveforms suggest a uterine myoma.

TVS [47] and transvaginal color Doppler imaging [48] were recently proposed for the diagnosis and follow-up of trophoblastic diseases and could replace pelvic arteriography. Myometrial invasion and image monitoring of the tumor response can be easily performed by TVS [47,48].

Umbilical cord abnormalities

The umbilical cord anatomy can be clearly identified early during pregnancy by color imaging (Figures 6.14 and 6.15).

The absence of one umbilical artery is amongst the most common congenital fetal malformations with an incidence of approximately 1% of all deliveries[49]. High resolution color Doppler imaging has an important role in early and accurate diagnosis of a single umbilical artery (Figure 6.19) and is of clinical value in view of the possible association of a single umbilical artery and intrauterine growth retardation with no major fetal malformation[49,50].

Cord angiomyxomas are vascular tumors in intimate contact with one or more umbilical vessels from which they probably arise, at both fetal or placental extremities of the cord[51,52]. Angiomas involving the placental insertion of the cord sometimes extend on to the fetal plate and may be more complicated to differentiate sonographically from a classical placental chorioangioma. Other cord tumors can also mimic a cord angioma on conventional ultrasound examination. Complete infiltration of the cord by the angiomatous tissue and secondary progressive shortening of the cord have serious implications for vaginal delivery[53]. With the knowledge of such a condition in the prenatal period, a Caesarean section can be recommended. Color flow Doppler imaging may also help the ultrasonographer (Figure 6.20) for early prenatal diagnosis of this type of cord abnormality[50].

REFERENCES

1. Ramsey, E.M. and Donner, N.W. (1980). *Placental Vasculature and Circulation.* (Stuttgart: Georg Thieme)
2. Hata, T., Hata, K., Senoh, D., Makihara, K., Aoki, S., Takamiya, O., Kitao, M. and Umaki, K. (1989). Transvaginal Doppler color flow mapping. *Gynecol. Obstet. Invest.*, **27**, 217–18
3. Wilkin, P. (1965). *Pathologie du Placenta.* (Paris: Masson)
4. Fox, H. (1978). *Pathology of the Placenta.* (Philadelphia: W.B. Saunders)
5. Moore, K.L. (1982). *The Developing Human: Clinically Oriented Embryology.* (Philadelphia: W.B. Saunders)
6. Itskovitz, J., Lindenbaum, E.S. and Brandes, J.M. (1980). Arterial anastomosis in the pregnant human uterus. *Obstet.Gynecol.*, **55**, 67–70
7. Brosens, I., Robertson, W.B. and Dixon, H.G. (1967). The physiological response of the vessels of the placental bed to normal pregnancy. *J. Pathol. Bact.*, **93**, 569–79
8. Robertson, W.B., Khong, M.B., Brosens, I., De Wolf, F., Sheppard, B.L. and Bonnar, J. (1986). The placental bed biopsy: review from three European centers. *Am. J. Obstet. Gynecol.*, **155**, 401–12
9. Hustin, J. and Schaaps, J.P. (1987). Echographic and anatomic studies of the maternotrophoblastic border during the first trimester of pregnancy. *Am.J. Obstet.Gynecol.*, **157**, 162–8
10. Hustin, J., Schaaps, J.P. and Lambotte, R. (1988). Anatomical studies of the utero-placental vascularization in the first trimester of pregnancy. *Trophoblast Res.*, **3**, 49–60
11. Schaaps, J.P. and Hustin, J. (1988). *In vivo* aspect of the maternal–trophoblastic border during the first trimester of gestation. *Trophoblast Res.*, **3**, 39–48
12. Schaaps, J.P. (1989). La grossesse jeune: mise a jour des malformations echographiques. *Contracep. Fertil. Sexual.*, **3**, 180–8
13. DeCherney, A.H., Romero, R. and Polan, M.L. (1982). Ultrasound in reproductive endocrinology. *Fertil. Steril.*, **37**, 323–8
14. Green, J.J. and Hobbins, J.C. (1988.) Abdominal ultrasound examination of the first-trimester fetus. *Am. J. Obstet. Gynecol.*, **159**, 165–75
15. Winberg, F. (1973). Echogenic changes with placental aging. *J. Clin. Ultrasound*, **1**, 52–5
16. Grannum, P.A.T., Berkowitz, R.L. and Hobbins, J.C. (1979). The ultrasonic changes in the maturing placenta and their relation to fetal pulmonic maturity. *Am. J. Obstet. Gynecol.*, **133**, 915–22
17. Fisher, C.C., Garrett, W. and Kossoff, G. (1976). Placenta aging monitored by gray scale echography. *Am. J. Obstet. Gynecol.*, **1**, 483–8
18. Haney, A.F. and Trought, W.S. (1980). The sonolucent placenta in high-risk obstetrics. *Obstet. Gynecol.*, **55**, 38–41
19. Vermeulen, R.C.W., Lambalk, N.B., Exalto, N. and Arts, N.F.T. (1985). An anatomic basis for ultrasound images of the human placenta. *Am. J. Obstet. Gynecol.*, 153, 806–10
20. Jauniaux, E. and Campbell, S. (1991). Perinatal assessment of placental and cord abnormalities. In Chervenak, F.A., Isaacson, G. and Campbell, S. (eds.) *The Textbook of Obstetrics and Gynecologic Ultrasound.* (Boston: Little, Brown and Co.) in press
21. Jauniaux, E., Avni, F.E., Elkazen, N., Wilkin, P. and Hustin, J. (1989). Etude morphologique des anomalies placentaires echographiques de la deuxieme moitie de la gestation. *J. Gynecol. Obstet. Biol. Reprod.*, **18**, 601–13
22. Jauniaux, E. and Campbell, S. (1990). Sonographic assessment of placental abnormalities. *Am. J. Obstet. Gynecol.*, in press
23. Kratochwill, A. and Eisenhut, L. (1967). Der fruheste Nachweis der fatalen Herzaction durch

Ultrascall. *Geburtsh. Frauenheilk.*, **27**, 176–80

24. Bernaschek, G., Rudelstorfer, R. and Csaicsich, P. (1988). Vaginal sonography versus serum human chorionic gonadotropin in early detection of pregnancy. *Am. J. Obstet. Gynecol.*, **158**, 608–12
25. Goldstein, S.R., Snyder, J.R., Watson, C. and Danon, M. (1988). Very early pregnancy detection with endovaginal ultrasound. *Obstet. Gynecol.*, **72**, 200–4
26. Fossum, G.T., Davajan, V. and Kletzky, O.A. (1988). Early detection of pregnancy with transvaginal ultrasound. *Fertil. Steril.*, **49**, 788–91
27. Enk, L., Wikland, M., Hammarberg, K. and Lindblom, B. (1990). The value of endovaginal sonography and urinary human chorionic gonadotropin tests for differentiation between intrauterine and ectopic pregnancy. *J. Clin. Ultrasound*, **18**, 73–8
28. Shapiro, B.S., Cullen, M., Taylor, K.J.W. and DeCherney, A.H. (1988). Transvaginal ultrasonography for the diagnosis of ectopic pregnancy. *Fertil. Steril.*, **50**, 425–9
29. McAnulty, J.H., Metcalfe, J. and Veland, K. (1982). Cardiovascular disease. In Burrow, G.N. and Ferris, T.F. (eds.) *Medical Complications during Pregnancy*, pp. 145–68. (Philadelphia: Saunders)
30. Silver, M., Barnes, R.J., Comline, R.S. and Burton, G.J. (1982). Placental blood flow: some fetal and maternal cardiovascular adjustments during gestation. *J. Reprod. Fertil.*, **31**, 139–45
31. Soothill, P.W., Nicolaides, K.H., Rodeck, C.H. and Campbell, S. (1986). Effect of gestational age and acid-base values in human pregnancy. *Fetal Ther.*, **1**, 168–75
32. Kaufmann, P. (1985). Influence of ischemia and artificial perfusion on placental ultrastructure and morphometry. *Contr. Gynecol. Obstet.*, **13**, 18–26
33. Burton, G.J., Ingram, S.C. and Palmer, M.E. (1987). The influence of mode of fixation on morphometrical data derived from terminal villi in the human placenta at term: a comparison of immersion and perfusion fixation. *Placenta*, **8**, 37–51
34. Farine, D., Fox, H.E., Jakobson, S. and Timor-Tritsch, I.E. (1988). Vaginal ultrasound for diagnosis of placenta previa. *Am. J. Obstet. Gynecol.*, **159**, 566–9
35. Farine, D., Fox, H.E., Jakobson, S. and Timor-Tritsch, I.E. (1989). Is it really placenta previa? *Eur. J. Obstet. Gynecol. Reprod. Biol.*, **31**, 103–8
36. Kurjak, A., Breyer, B., Jurković, D., Alfirevic, Z. and Miljan, M. (1987). Color flow mapping in obstetrics. *J. Perinat. Med.*, **15**, 271–81
37. Kurjak, A., Alfirevic, Z. and Miljan, M. (1988). Conventional and color Doppler in the assessment of fetal and maternal circulation. *Ultrasound Med. Biol.*, **14**, 337–54
38. Hustin, J., Foidart, J.M. and Lambotte, R. (1983). Maternal vascular lesions in pre-eclampsia and intrauterine growth retardation: light microscopy and immunofluorescence. *Placenta*, **4**, 489–98
39. Labarrere, C.A. (1988). Acute atherosis. An histopathological hallmark of immune aggression? *Placenta*, **9**, 95–108
40. Labarrere, C.A., Althabe, O., Caletti, E. and Muscolo, D. (1986). Deficiency of blocking factors in intrauterine growth retardation and its relationship with chronic villitis. *Am. J. Reprod. Immun. Microbiol.*, **10**, 14–19
41. Gilles, W.B., Trudinger, B.J. and Baird, P.J. (1985). Fetal umbilical artery flow velocity waveforms and placental resistance: pathological correlation. *Br. J. Obstet. Gynaecol.*, **92**, 31–8
42. McCowan, L.M., Mullen, B.M. and Ritchie, K. (1987). Umbilical artery flow velocity waveforms and the placental vascular bed. *Am. J. Obstet. Gynecol.*, **157**, 900–2
43. Jimenez, E., Vogel, M., Arabin, B., Wagner, G. and Mirsalim, P. (1988). Correlation of ultrasonographic measurements of the uteroplacental and fetal blood with the morphological diagnosis of placental function. *Trophoblast Res.*, **3**, 325–34
44. Fok, R.Y., Pavlova, Z., Benirschke, K., Paul, R.H. and Platt, L.D. (1990). The correlation of arterial lesions with umbilical artery Doppler velocimetry in the placentas of small-for-dates pregnancies. *Obstet. Gynecol.*, **75**, 578–83
45. Nessmann, C., Huten, Y. and Uzan, M. (1988). Placental correlations of abnormal umbilical Doppler index. *Trophoblast Res.*, **3**, 309–24
46. Jauniaux, E. and Campbell, S. (1990). Fetal growth retardation with abnormal blood flows and placental sonographic lesions. *J. Clin. Ultrasound*, **18**, 210–14
47. Schneider, D.F., Bukovsky, I., Weinraub, Z., Golan, A. and Caspi, E. (1990). Transvaginal ultrasound diagnosis and treatment follow-up of invasive gestational trophoblastic disease. *J. Clin. Ultrasound*, **18**, 110–13
48. Aoki, S., Hata, T., Hata, K., Senoh, D., Miyako, J., Takamiya, O., Iwanari, O. and Kitao, M. (1989). Doppler color flow mapping of an invasive mole. *Gynecol. Obstet. Invest.*, **27**, 52–4
49. Jauniaux, E., De Munter, C., Pardou, A., Elkhazen, N., Rodesch, F. and Wilkin, P. (1989). Evaluation echographique du syndrome de l'artere ombilicale unique: une serie de 80 cas. *J. Gynecol. Obstet. Biol. Reprod.*, **18**, 341–8
50. Jauniaux, E., Campbell, S. and Vyas, S. (1989). The use of color Doppler imaging for prenatal diagnosis of umbilical cord anomalies: report of three cases. *Am. J. Obstet. Gynecol.*, **161**, 1195–7
51. Heifetz, S.A. and Rueda-Pedraya, M.E. (1983). Hemangiomas of the umbilical cord. *Pediatr. Pathol.*, **1**, 385–93
52. Jauniaux, E., De Munter, C., Vanesse, M., Hustin, J. and Wilkin, P. (1989). Embryonic remnants of the umbilical cord: morphological and clinical aspects. *Hum. Pathol.*, **20**, 458–62
53. Jauniaux, E., Moscoso, G., Chitty, L., Gibb, D., Driver, M. and Campbell, S. (1990). An angiomyxoma involving the whole length of the umbilical cord: prenatal diagnosis by ultrasonography. *J. Ultrasound Med.*, **9**, 419–22

7 Embryonic and Fetal Circulation

M. Miljan and A. Kurjak

Ultrasound imaging has established its pre-eminence in non-invasive antenatal diagnosis by depicting anatomical change, while the Doppler technique offers the potential to study functional and hence physiological changes. Recently, a new diagnostic method, transvaginal color Doppler, was introduced, enabling simultaneous visualization of small embryonic structures as well as blood flow in embryonic vessels.

EMBRYOLOGIC CONSIDERATIONS

The vascular system of the human embryo appears in the middle of the 3rd week, when the embryo is no longer able to satisfy its nutritional requirements by diffusion alone[1]. The occurrence of the primitive circulation is also well correlated with the enlargement of the chorionic vesicle[2]. About 13 days after the commencement of implantation (20-day-old embryo or 34 days from last menstrual period), a primitive circulatory system is formed in the embryo, chorion, yolk sac and connective stalk. In the meantime, complex changes appear leading to the transformation of the primitive embryonic cardiovascular system into the well-established cardiovascular system as seen in the fetus. Another set of changes, the definitive one, will finally occur after birth, adapting the fetal circulation to the new conditions of life.

The entire cardiovascular system, the heart, blood vessels and blood cells, originates from the mesodermal germ layer. Mesenchymal cells in the splanchnic mesoderm layer of the late presomite embryo proliferate and form isolated cell clusters known as the angiogenetic clusters. With time they acquire a lumen, unite, and form a plexus of small blood vessels. The anterior central portion of this plexus, known as the cardiogenic area, gives origin to the paired tubes which will form the fetal heart. Other clusters, located bilaterally, parallel and close to the mid-line of the embryonic shield, also acquire a lumen and form a pair of longitudinal vessels, dorsal aortae.

By the 21st day postimplantation (5 weeks of gestational age, embryonal length 1.5 mm), the embryonal circulation as well as the circulation from the embryo through the blood vessels is established and the fetal heart is pulsating.

Heart formation

Although initially paired, by the 22nd day of development the two heart tubes form a single, slightly bent heart tube. During the 4th to the 7th week, the heart is partitioning and differentiating, thus becoming a typical four-chambered structure.

The arterial system

The destiny of the embryonic arterial system is closely related to the destiny of the branchial arches. When the branchial arches are formed during the 4th and 5th week of development, each arch (six of them) receives its own cranial nerve and artery (aortic arches). Aortic arches arise from the aortic sac, the most distal part of the heart tube, connecting it with the dorsal aorta. Later on, most of the aortic arches obliterate entirely or partially, giving rise to the definitive pattern of the embryonic arterial system.

The venous system

The embryonic venous system has its origin in three primitive venous complexes. In the 5th week three pairs of major veins can be distinguished: vitelline or omphalomesenteric, umbilical, and cardinal veins. After a complex transformation, further main embryonic veins and vein systems may be found: the portal system, which originates from vitelline veins; the caval system, which originates from cardinal veins; and finally the umbilical venous system, a part of which is the ductus venosus.

Therefore, the whole embryonic cardiovascular system is almost completely formed by the 7th developmental week.

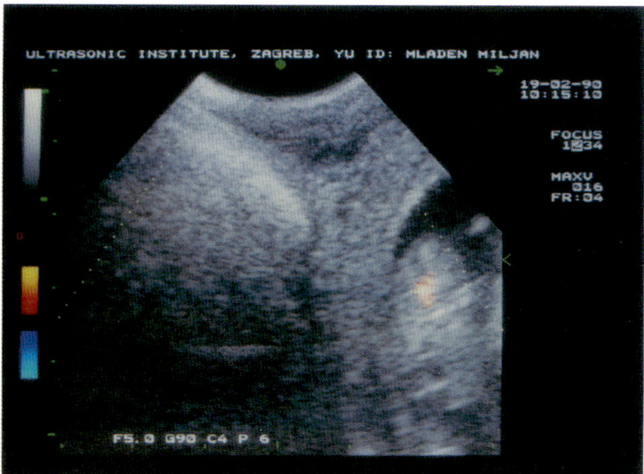

Figure 7.1 Color Doppler coded intracardiac blood flow (red) in embryo aged 8 weeks

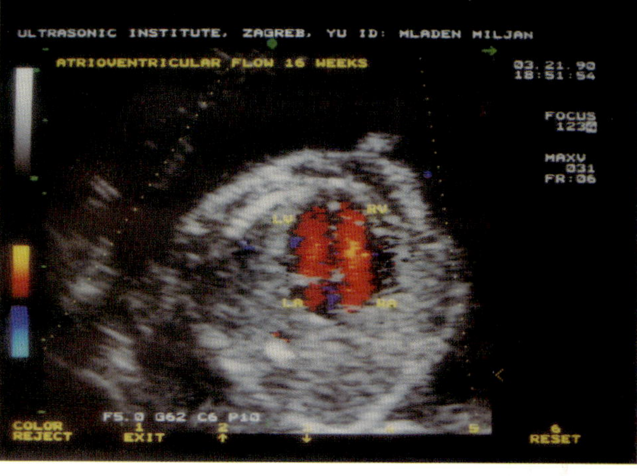

Figure 7.2 Color Doppler superimposed on the four-chamber view of the fetal heart in 16th week, obtained by transvaginal approach. Atrioventricular flow (red) is detected during ventricular diastole. LA = left atrium; RA = right atrium; LV = left ventricle; RV = right ventricle

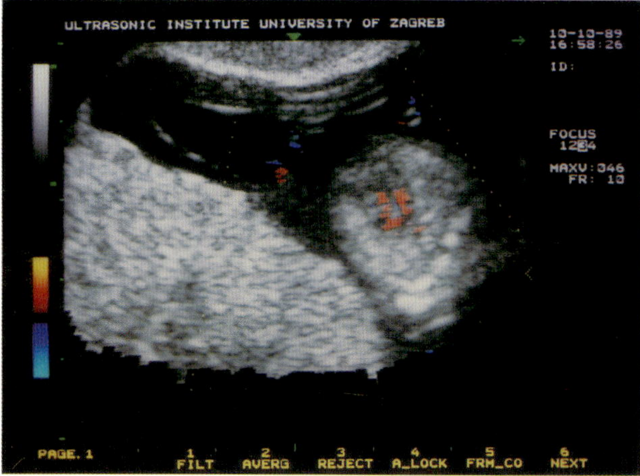

Figure 7.3 Color Doppler superimposed on the four-chamber view of the fetal heart in 16th week, obtained by transabdominal approach

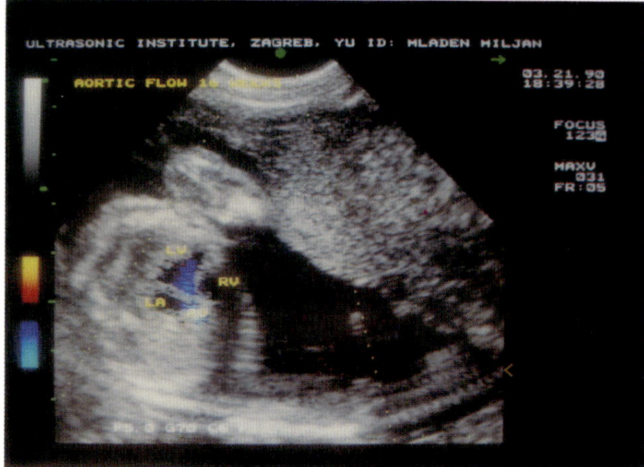

Figure 7.4 Aortic blood flow (blue) detected by color Doppler on the long axis of the left ventricle. LA = left atrium; LV = left ventricle; AO = ascending aorta; RV = right ventricle

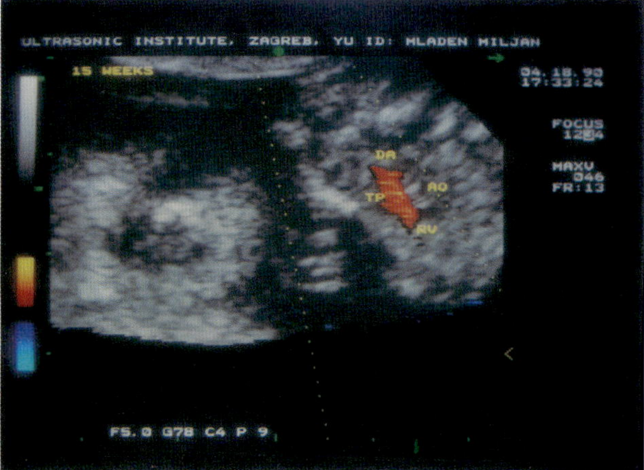

Figure 7.5 Color Doppler superimposed on the long axis of the right ventricle (RV) enables visualization of blood flow through the pulmonary trunk (TP) (red). DA = ductus arteriosus

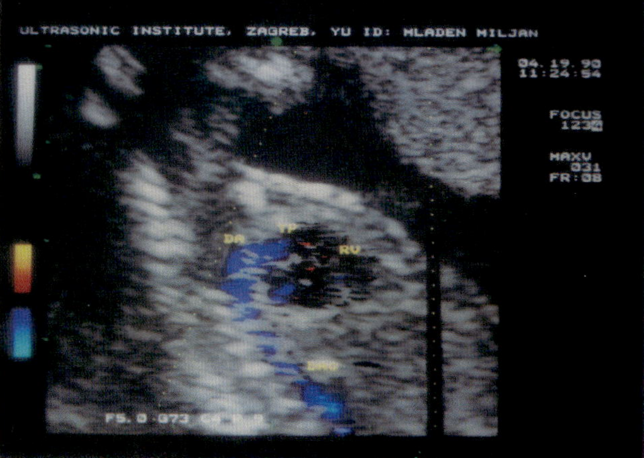

Figure 7.6 Color-coded blood flow through the pulmonary trunk (TP), ductus arteriosus (DA) and descending aorta (DAO) (blue). RV = right ventricle

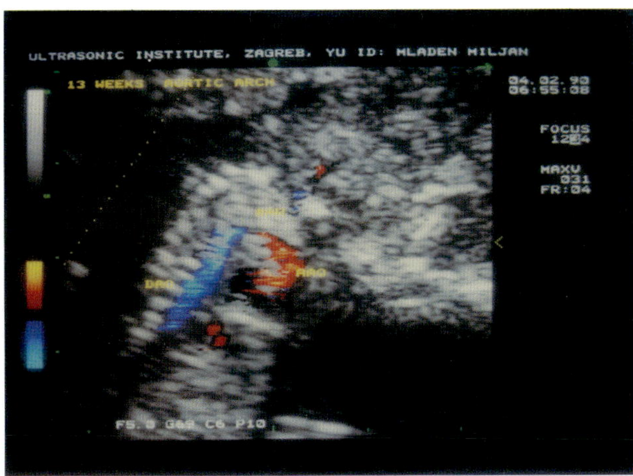

Figure 7.7 Sagittal section through the fetal thorax demonstrating flow through the ascending aorta (AAO), aortic arch (AAR) (red) and descending aorta (DAO) (blue)

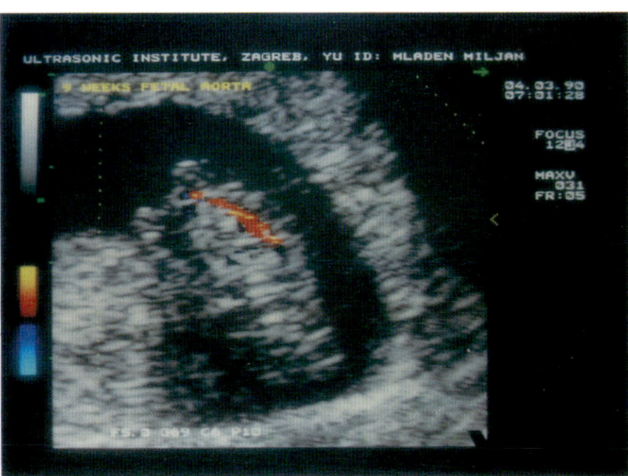

Figure 7.8 Color-coded blood flow in the abdominal part of the descending aorta in 9-week fetus (red)

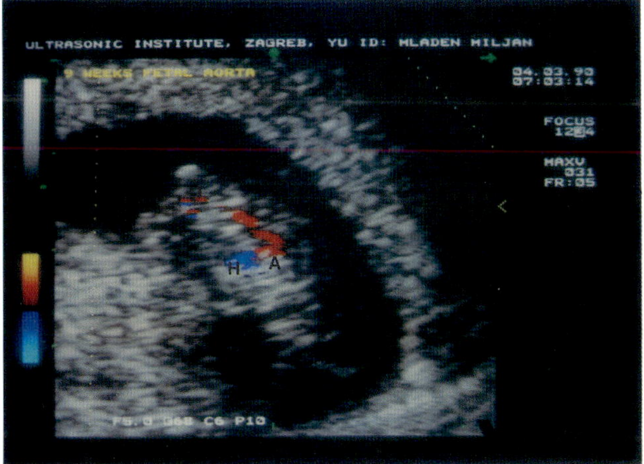

Figure 7.9 Sagittal section of the same fetus shown in Figure 7.8, demonstrating blood flow in the fetal heart (H) (blue), aortic arch (A) and descending aorta (red), as well as in the iliac arteries (I) (red)

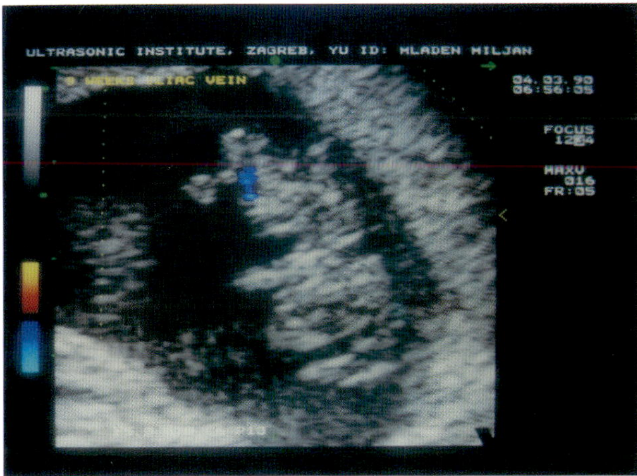

Figure 7.10 Blood flow (blue) in the iliac veins in the same fetus as shown in Figures 7.8 and 7.9

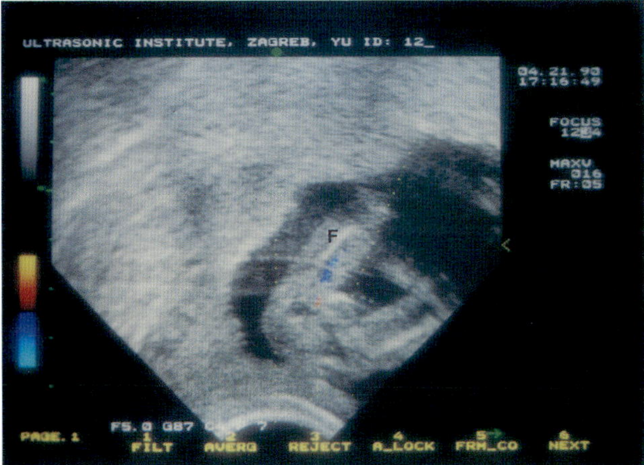

Figure 7.11 Blood flow in femoral vessels (red, artery; blue, vein) in a fetus aged 15 weeks. F = fetal femur

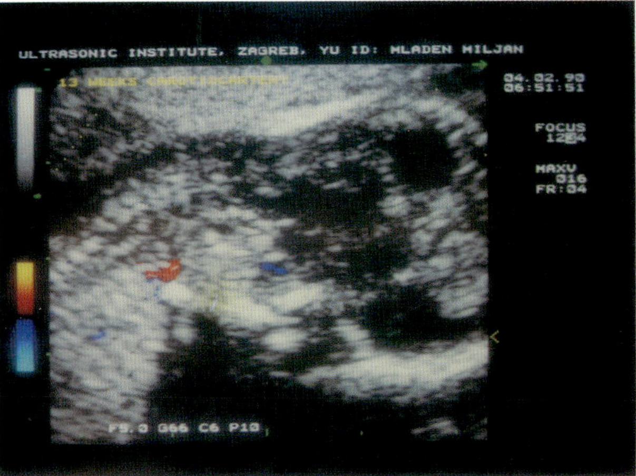

Figure 7.12 Parasagittal section through the neck of a 13-week-old fetus demonstrating blood flow through the carotid artery (red)

CLINICAL FINDINGS

The embryonal circulation, or, to be precise, a part of it, becomes visible in a very early stage of pregnancy. While transabdominal examination enables detection of embryonic heart beats from the 7th week, the transvaginal approach offers this possibility much earlier. The fetal pole becomes evident by transvaginal sonography at the end of the 5th and during the 6th week, and at the same time fetal heart beats are seen[3,4]. The intracardiac flow, detectable in this period, is the first embryonic flow visualized by color Doppler. Later on, detection of blood flow in particular parts of the fetal heart, as well as in major embryonic vessels, becomes possible. A comparison between the results which can be obtained using the transabdominal versus the transvaginal approach, as well as the possibilities of transvaginal Doppler techniques in the evaluation of embryonic and fetal blood flow, will be presented.

THE HEART AND GREAT VESSELS

The first embryonic structure in which blood flow can be detected by color Doppler is precisely the embryonic heart. The earliest detection of embryonic heart beats is possible at 5 weeks and 4 days (25 days postconception), i.e. when the embryo becomes visible. During the 6th week heart beats have been detected in 94% of examined cases[5]. At the same time, intracardiac flow can be visualized employing the color Doppler technique[6]. Initially, it appears as a pulsating point inside the embryonic pole, which becomes more prominent and easily detectable later on during the first trimester (Figure 7.1). Meanwhile, as a clear two-dimensional visualization of cardiac structures is not possible before 12 weeks, a precise localization of intracardiac flow inside the cardiac chambers and great vessels during the first trimester is somewhat limited.

During the past 7 years fetal echocardiography has progressed from tentative beginnings into a sufficiently reliable technique. Using high-resolution transabdominal probes, fetal heart structures could be visualized as early as the 16th week of pregnancy[7], which enabled antenatal detection of congenital heart defects[8–10]. After successful use in pediatric and adult echocardiography, color Doppler was also introduced in fetal echocardiography[1–14]. Following the development of high-resolution high-frequency transvaginal probes, interest has been focused on the possibility of utilizing this improved technique to define normal and abnormal morphology and the physiology of the human fetal heart in the late first and early second trimesters. Although some cardiac structures, such as the interventricular septum, can be detected as early as the 10th week of pregnancy[5], detailed evaluation of cardiac morphology becomes possible a few weeks later. At the end of the 11th week of pregnancy, the fetal heart resembles a trilocular cavity[15]. Both ventricular chambers can be distinguished, divided by a well-defined interventricular septum, but at that gestational age the interatrial septum with the foramen ovale cannot be visualized. It is well known that the interatrial septum is almost completely formed (except for the final closure of the foramen ovale which happens after delivery) by the 55th postconceptual day[16]. Therefore, it is believed that failure to demonstrate the interatrial septum at the 11th week of pregnancy is not caused by its absence, but by the limited possibilities of ultrasound equipment even when high-resolution high-frequency transvaginal probes are used. Therefore, diagnosis of an atrial septal defect prior to the 12th week of pregnancy is uncertain when only the two-dimensional technique without color Doppler is used. Despite some limitations of two-dimensional echocardiography in the late first trimester of pregnancy, recent reports[6,15] suggest that screening for structural anomalies can already be initiated at the early second trimester of pregnancy.

From the 12th week of gestation, when cardiac structures are clearly visible, simultaneous echocardiographic evaluation of cardiac morphology and intracardiac blood flow becomes possible using two-dimensional Doppler color flow mapping[5,6,15]. For this purpose, the same echocardiographic planes are used as in the transabdominal approach. Thus, a four-chamber view allows visualization of atrioventricular flow (Figures 7.2 and 7.3). The inflow and outflow of the left ventricle may be studied using the long-axis view of this ventricle (Figure 7.4), while the long axis of the right ventricle allows detection of right ventricular outflow (Figure 7.5). The short axis through the basis of the heart at the level of the great vessels enables simultaneous visualization of the right ventricular inflow and outflow, and, by slight angulation of the transducer, the ductus arteriosus plane is obtained, demonstrating blood flow from the pulmonary truncus, through the ductus arteriosus, into the descending aorta (Figure 7.6). On the sagittal section through the fetal thorax, aortic arch flow is visualized, and, if the fetus is lying in a favorable position, the flows in both the ascending and descending aorta are simultaneously detected (Figure 7.7).

Visualization of intracardiac blood flow in the early second trimester of pregnancy varies throughout gestation. The lowest success rate was found in the 12th week (78.5%), rising to 90.5% and 93.5% in the 13th and 14th weeks, respectively, while in the 15th and 16th weeks complete evaluation of intracardiac flow was possible in over 95% of examined fetuses[6].

A number of factors may influence our ability to demonstrate intracardiac and great vessel blood flow. Among them the following were found to be the most important: dimensions of the heart structures, distance between the fetal heart and the transducer, fetal position and excessive fetal movements. Nevertheless, transvaginal color Doppler represents a useful innovation in fetal echocardiography because it offers the possibility of an accurate evaluation of cardiac and great vessel blood flow much earlier than was possible with the transabdominal approach. According to our experience from transabdominal color Doppler echocardiography, one may suppose that transvaginal color Doppler assessment of intracardiac hemodynamics will enable very early detection of fetal cardiac valve disorders such as valvular insufficiency.

The pulse Doppler echocardiographic technique has allowed quantification of cardiac blood flow patterns from the 18th week to term[17–22]. Despite various factors limiting transvaginal echocardiographic examination, initial attempts are also being made to quantify intracardiac blood flow in early pregnancy. Pulse Doppler measurements performed in 30 fetuses from 11 to 13 weeks showed significant differences between some cardiac flow patterns (peak systolic velocity, time average velocity) in early gestation compared to that found in late second and third trimester pregnancies, exhibiting lower ventricular compliance and high cardiac afterload in early second trimester fetuses[23].

Up to now results obtained by transvaginal echocardiography have been promising and encouraging, and it is to be expected that further clinical work will highlight its potential in the early evaluation of fetal cardiac function.

THE AORTA AND MAIN ARTERIES

The aorta is the second structure in which blood flow is recognizable during the early embryonic period. It becomes possible from the 8th week of pregnancy. In the beginning, the descending part of the aorta is usually visualized, but 1 week later (9th week) the aortic arch (Figures 7.8 and 7.9) and flow through the iliac vessels (Figures 7.9 and 7.10) are detectable. In the early second trimester, when the fetus lies in a convenient position, even blood flow through the femoral vessels can be detected (Figure 7.11).

Blood flow through the main aortic branches which supply the head and neck becomes detectable in the early second trimester. Thus, carotid artery flow becomes evident during the 13th week of pregnancy (Figure 7.12). Simultaneously, close to the carotid artery, the jugular vein and its flow may be detected (Figure 7.13).

Recent years have witnessed a surge of interest in the application of Doppler ultrasound velocimetry as a fetal diagnostic tool. Among different fetal and materno–placental vessels investigated by the transabdominal probe during the late second and third trimesters, the fetal aorta has been under investigation[24, 25]. The transvaginal approach enables measurement of flow patterns in the fetal aorta much earlier than was possible by using the transabdominal probe. Preliminary data obtained on 30 fetuses between 11 and 13 weeks gestation demonstrated the pulsatility index to be about 2.5. The estimated peak systolic Doppler shift in this group of fetuses was found to be about 4 cm/s, and shows increment values with advanced gestation, while time average velocity in that period ranged between 15 and 18 cm/s[23]. The same authors gave some preliminary results about flow patterns in the common carotid artery. The pulsatility index ranged between 2.8 and 3, the peak systolic velocity was found to be between 2 and 2.5 cm/s, while the time average velocity was about 8 cm/s[23].

INTRACRANIAL CIRCULATION

Recently, a pulsed Doppler method was introduced for recording the flow velocity waveforms in the fetal internal carotid artery[26, 27], in the middle[28, 29] and in the anterior and posterior cerebral arteries[30] during late pregnancy, and direct evidence was provided for the occurrence of a brain-sparing effect in cases of intrauterine growth retardation[27, 31].

Transvaginal sonography enabled visualization of intracranial structures earlier than was possible by the transabdominal approach. Thus, at 8 weeks of gestation, the first documentation of ventricles is provided, while a week later the right and left ventricles, with a choroid plexus, can be seen separately[3, 4]. The intracranial circulation becomes visible at the 10th week (Figure 7.14). At that time, discrete pulsations of intracranial parts of the internal carotid arteries may be detected at the base of the skull. Blood flow in the cerebral arteries can be clearly distinguished 1 week later. We found that the horizontal plane through the head at the level of temporal and sphenoidal bones is more convenient for visualizing flow through the middle and posterior cerebral arteries (Figures 7.15 and 7.16), while the coronal plane enables easier and complete visualization of the anterior cerebral artery as well as its origin from the internal carotid artery (Figure 7.17). Using a horizontal plane at the level of the sphenoidal and temporal bones, complex communication between the basilar and internal carotid artery, named the circle of Willis, can be detected (Figure 7.18). Preliminary results obtained

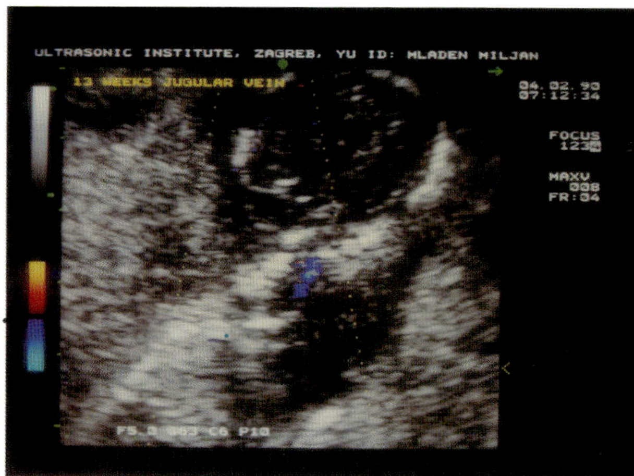

Figure 7.13 Jugular blood flow (blue) in the same fetus as demonstrated in Figure 7.12

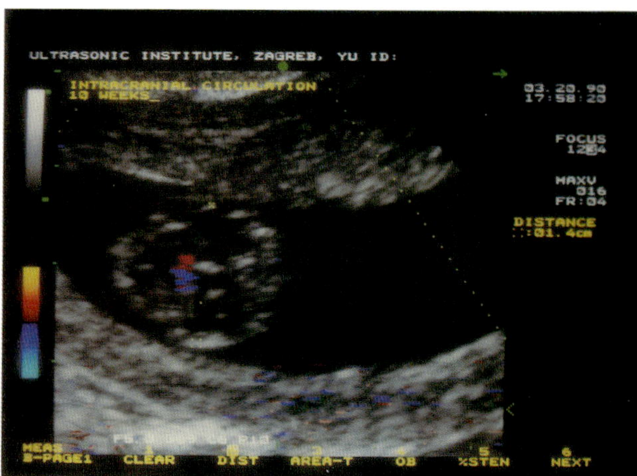

Figure 7.14 Intracranial blood flow in a 10-week-old fetus (red–blue)

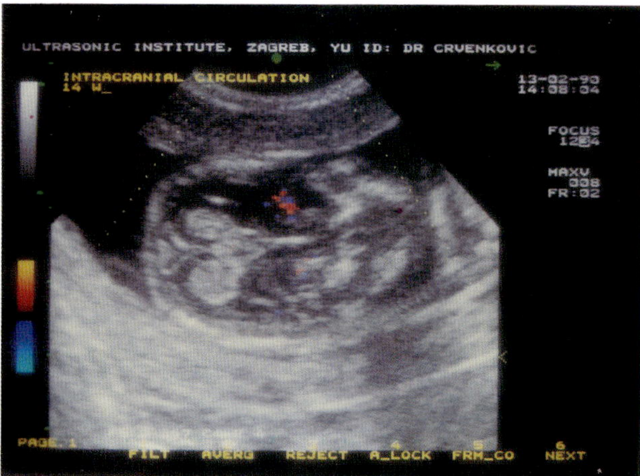

Figure 7.15 Blood flow through middle cerebral artery (red) in a 14-week-old fetus

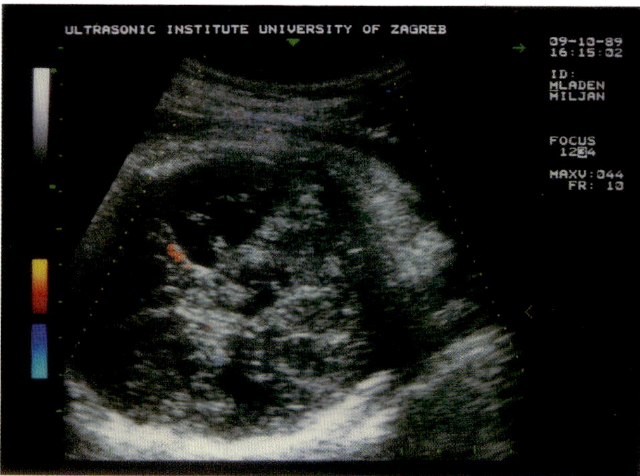

Figure 7.16 Blood flow through the posterior cerebral artery (red) in a 16-week-old fetus

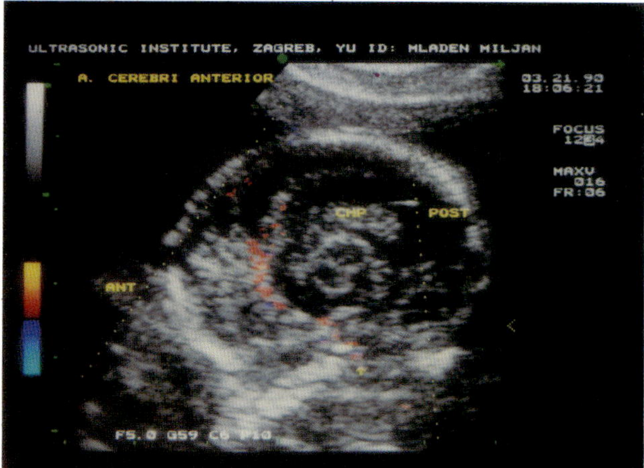

Figure 7.17 Color Doppler superimposed on sagittal section through the fetal head demonstrating blood flow in the anterior (ANT) cerebral artery (red). Arrow indicates blood flow in internal carotid artery at point of entry into cranial cavity. CHP = choroid plexus; POST = posterior

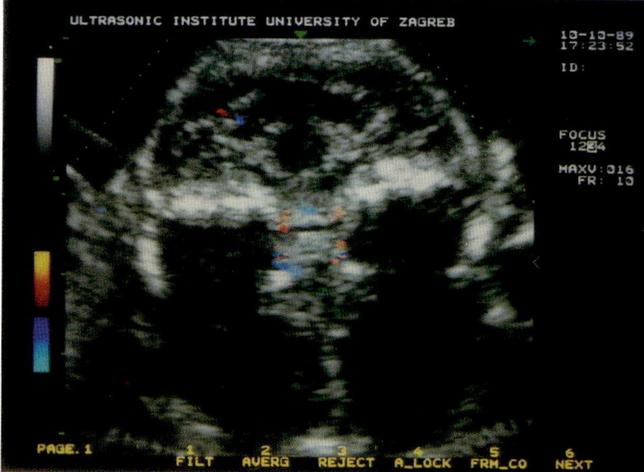

Figure 7.18 Horizontal plane through fetal skull at the level of sphenoidal and temporal bones, demonstrating blood flow through the circle of Willis (red–blue)

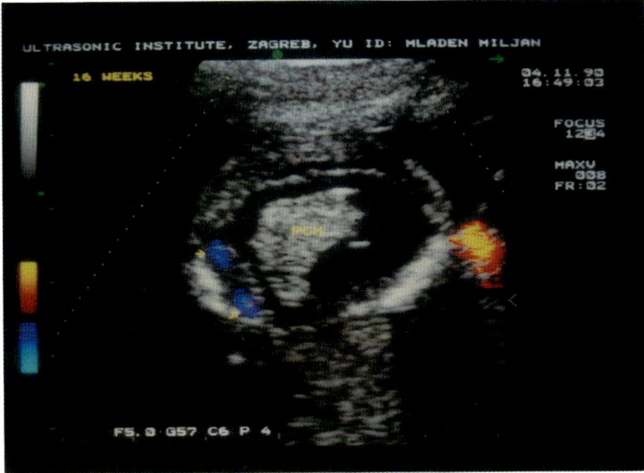

Figure 7.19 Color Doppler superimposed on the sagittal section through 16-week-old fetus, demonstrating blood flow in the sinus sagittalis superior (blue, indicated by arrows). PCH = choroid plexus

from fetal intracerebral arteries between 11 and 13 weeks of pregnancy showed that the pulsatility index is significantly higher compared to that found in second and third trimester fetuses, and it gradually decreases during pregnancy[23].

Besides the arterial vessels, a part of the venous intracranial flow may also be studied. From the complex intracranial venous drainage we were able to visualize flow in the superior sagittal sinus, which is located posteriorly, just below the occipital bone (Figure 7.19).

CONCLUSION

Transvaginal sonography provides a closer look at the embryonic and early fetal development through more detailed imaging of different embryonic and fetal structures. With the introduction of the color Doppler technique, not only is embryonic and fetal morphology visualized, but detailed evaluation of embryonic and fetal circulation becomes feasible. Experience with the use of color Doppler transvaginal sonography in evaluating embryonic and fetal circulation is still in its formative stages, but even now some important advantages of this method can be foreseen. Thus, for example, the evaluation of fetal heart morphology and intracardiac blood flow becomes available 2–4 weeks earlier than by the transabdominal approach. This may provide revolutionary changes in fetal echocardiographic examinations. It is quite impossible at this moment to imagine all the possibilities which this new technique will offer in the future investigation of fetal well-being, but, based upon our experience, the benefits are more than obvious.

REFERENCES

1. Langman, J. (1981). *Medical Embryology*, 4th edn., p. 157. (Baltimore, London: Williams & Wilkins)
2. Hamilton, W.J., Boyd, J.D. and Mossman, H.W. (1945). The implantation of the blastocyst and the development of the fetal membranes, placenta and decidua. In *Human Embryology*, p. 49. (Cambridge: W. Heffer & Sons Ltd.)
3. Blumenfeld, Z., Rottem, S., Elgali, S. and Timor-Tritsch, I.E. (1987). Transvaginal sonographic assessment of early embryological development. In Timor-Tritsch, I.E. and Rottem, S. (eds.) *Transvaginal Sonography*, p. 87. (New York: Elsevier)
4. Timor-Tritsch, I.E., Rottem, S. and Thaler, I. (1988). Review of transvaginal ultrasonography: description with clinical application. *Ultrasound Q.*, **6**, 1
5. Timor-Tritsch, I.E., Peisner, D.B. and Raju, S. (1990). Sonoembriology: an organ-oriented approach using a high-frequency vaginal probe. *J. Clin. Ultrasound*, **18**, 286
6. Miljan, M. and Kurjak, A. (1990). Fetal color Doppler echocardiography in the first and early second trimester of pregnancy. *EUROSON 90, 7th Congress of the European Federation of Societies for Ultrasound in Medicine and Biology*, Jerusalem, May 6–11, Abstr. p. 93
7. Allan, L.D., Tynan, M., Campbell, S., Wilkinson, J.L. and Anderson, R.H. (1980). Echocardiographic and anatomical correlates in the fetus. *Br. Heart J.*, **44**, 444
8. Kleinman, C.S., Hobbins, J.C., Jaffe, C.C., Lynch, D.C. and Talner, N.S. (1980). Echocardiographic studies on the human fetus: prenatal diagnosis of congenital heart disease and cardiac dysrhythmias. *Pediatrics*, **65**, 1059
9. Allan, L.D., Crawford, D.C., Anderson, R.H. and Tynan, M.J. (1984). Echocardiographic and anatomical correlations in fetal congenital heart disease. *Br. Heart J.*, **52**, 542
10. Wladimiroff, J.W., Stewart, P.A. and Tonge, H.M. (1984). The role of diagnostic ultrasound in the study of fetal cardiac abnormalities. *Ultrasound Med. Biol.*, **10**, 457
11. DeVore, G.R., Horenstein, J., Siassi, B. and Platt, L.D. (1987). Doppler color flow mapping: a new technique for the diagnosis of congenital heart defects. *Am. J. Obstet. Gynecol.*, **156**, 1054
12. Kurjak, A., Breyer, B., Jurković, D., Alfirevic, Z. and Miljan, M. (1987). Color flow mapping in obstetrics. *J. Perinat. Med.*, **15**, 271
13. Kurjak, A., Alfirevic, Z. and Miljan, M. (1988). Conventional and color Doppler in the assessment of fetal and maternal circulation. *Ultrasound Med. Biol.*, **14**, 337
14. Kurjak, A., Miljan, M., Jurković, D., Alfirevic, Z. and Žalud, I. (1989). Color Doppler in the assessment of fetomaternal circulation. *Rech. Gynecol.*, **1**, 269
15. Rottem, S. and Bronshtein, M. (1990). Transvaginal sonographic diagnosis of congenital anomalies between 9 weeks and 16 weeks, menstrual age. *J. Clin. Ultrasound* , **18**, 307
16. Nora, J.J. and Nora, A.H. (1978). *Genetics and Counseling in Cardiovascular Diseases*. (Springfield, Illinois: Charles C. Thomas)
17. Maulik, D., Nanda, N.C. and Saini, V.D. (1984). Fetal Doppler echocardiography: methods and characterization of normal and abnormal hemodynamics. *Am. J. Cardiol.*, **53**, 572

18. Reed, K.L., Meijboom, E.J., Sahn, D.J., Scagnelli, S.A., Valdes-Cruz, L.M. and Shenker, L. (1986). Cardiac Doppler flow velocities in human fetuses. *Circulation*, **73**, 41
19. Allan, L.D., Chita, S.K., Al-Ghazali, W., Crawford, D. and Tynan, M. (1987). Doppler echocardiographic evaluation of the normal human fetal heart. *Br. Heart J.*, **57**, 528
20. Machado, M.V.L., Chita, S.C. and Allan, L.D. (1987). Acceleration time in the aorta and pulmonary artery measured by Doppler echocardiography in the midtrimester fetus. *Br. Heart J.*, **58**, 15
21. De Smedt, M.C.H., Visser, G.H.A. and Meijboom, E.J. (1987). Fetal cardiac output estimated by Doppler echocardiography during mid- and late gestation. *Am. J. Cardiol.*, **60**, 338
22. Reed, K.L., Anderson, C.F. and Shenker, L. (1987). Fetal pulmonary artery and aorta: two-dimensional Doppler echocardiography. *Obstet. Gynecol.*, **69**, 175
23. Stewart, P.A. and Wladimiroff, J.W. (1990). Fetal blood flow studies in IUGR. *EUROSON 90, 7th Congress of the European Federation of Societies for Ultrasound in Medicine and Biology*, Jerusalem, May 6–11, Abstr. p. 81
24. Eldridge, M.W. (1981). Doppler ultrasound measurements of human fetal aortic blood flow. *Biomed. Sci. Instrum.*, **17**, 67
25. Jouppila, P. and Kirkinen, P. (1984). Increased vascular resistance in the descending aorta of the human fetus in hypoxia. *Br. J. Obstet. Gynaecol.*, **91**, 863
26. Wladimiroff, J.W., Tonge, H.M. and Stewart, P.A. (1986). Doppler ultrasound assessment of cerebral blood flow in the human fetus. *Br. J. Obstet. Gynaecol.*, **93**, 471
27. Wladimiroff, J.W., Van Den Wijngaard J.A.G.W., Degani, S., Noordom, M.J., Van Eyck, J. and Tonge, H.M. (1987). Cerebral and umbilical arterial blood flow velocity waveforms in normal and growth-retarded fetuses. *Obstet. Gynecol.*, **69**, 705
28. Woo, J.S.K., Liang, S.T., Lo, R.L.S. and Chan, F.Y. (1987). Middle cerebral artery Doppler flow velocity waveforms. *Obstet. Gynecol.*, **70**, 613
29. Veile, J-C. and Cohen, I. (1990). Middle cerebral artery blood flow in normal and growth-retarded fetuses. *Am. J. Obstet. Gynecol.*, **162**, 391
30. Van Den Wijngaard, J.A.G.W., Groenenberg, I.A.L., Wladimiroff, J.W. and Hop, W.C.J. (1989). Cerebral Doppler ultrasound of the human fetus. *Br. J. Obstet. Gynaecol.*, **96**, 845
31. Arduini, D., Rizzo, G. and Romanini, C. (1987). Fetal blood flow velocity waveforms as predictors of growth retardation. *Obstet. Gynecol.*, **70**, 7

8 The Second and Third Trimesters of Pregnancy

A. Kurjak and M. Miljan

During the past few years the pulse Doppler technique enabled quantification of blood flow patterns in various fetal vessels. The use of this technique was first reported in 1977, when FitzGerald and Drumm[1] demonstrated Doppler frequency shift waveforms from the umbilical arterial circulation. Since then, much data have been collected demonstrating the feasibility of its use for investigating the various components of the fetal circulation: the umbilical vein, aorta[2, 3], heart[4–9] and cerebral[10–14] circulation. The duplex scanner systems, a combination of real-time imaging and Doppler technique, have improved antenatal measurements by allowing precise positioning and permanent control of the sample volume gate. This system, meanwhile, is useful only in cases where the dimension of the vessel of interest is large enough that it can be visualized by the two-dimensional technique. The problem arises when one wants to detect blood flow in those small vessels whose width is under the detection possibilities of the two-dimensional technique. Recently, color Doppler was introduced in obstetrics, allowing visualization of fetal, placental and maternal circulation[15–18].

As has been demonstrated in the previous chapters, embryonic, placental and uterine blood flows in the first and early second trimesters of pregnancy are better visualized by the transvaginal approach. An advance in gestational age is accompanied by a rapidly increasing volume of amniotic fluid and a fast-growing fetus. After the 16th week of pregnancy, the fetus overgrows the actual focal zone of the transvaginal probe, and only those structures which are nearer to the probe may be visualized. Therefore, the use of the transabdominal probe for the second and third trimester examinations seems to be more convenient.

Color Doppler has enabled visualization of almost the whole maternal, placental and fetal circulations. Detection of blood vessels by color Doppler compared to that obtained by the duplex system alone is easier, faster and more accurate: thus the time of investigation becomes shorter and the Doppler examination more secure. In this chapter, we will briefly review the use of color Doppler in the late second and third trimesters of pregnancy.

Maternal pelvic vessels can be easily detected by placing the transducer obliquely at the lateral uterine borders. Usually, the maternal iliac artery lies parallel to the uterine border (Figure 8.1), while the uterine artery crosses it (Figure 8.2).

The placenta is not completely formed until the second trimester of pregnancy. By the beginning of the fourth month, the placenta has two components: a fetal portion formed by the chorion frondosum, and a maternal portion formed by the decidua basalis. Between the chorionic plate, which represents the borderline from the fetal side, and the decidual plate, which represents the inner surface of the uterine wall, there are numerous intervillous spaces which are filled with maternal blood. The cotyledons receive their blood through numerous spiral arteries which pierce the decidual plate and enter the intervillous spaces. By color Doppler it is possible to visualize the blood flow through the intervillous spaces (Figure 8.3).

Blood from the placenta returns to the fetus by way of the umbilical vein. It can be visualized by color Doppler during its course through the umbilical cord (Figures 8.4 and 8.5), or inside the fetal abdomen (Figure 8.6). On approaching the liver, usually the main volume of the blood flows through the ductus arteriosus, short-circuiting the liver, and the smaller volume enters the liver and mixes with blood from the portal circulation. Blood flow through the ductus arteriosus can be easily detected by color Doppler by a sagittal section through the fetal abdomen, and its course from the umbilical vein to the inferior vena cava can be completely visualized (Figures 8.7 and 8.8). With the color Doppler equipment presently used, only a part of the complex circulation through the liver can be seen, i.e. the blood flow through the hepatic vessels (Figure 8.9). The whole blood which by-passes the liver collects into the main hepatic vein (Figure 8.10), and enters into the inferior vena cava (Figure 8.11). After a short course in the inferior vena cava, blood enters the right atrium (Figure 8.11).

The blood from the inferior vena cava is largely directed by the lower border of the septum secundum, the crista dividens, through the foramen ovale into the left atrium (Figure 8.12). A small

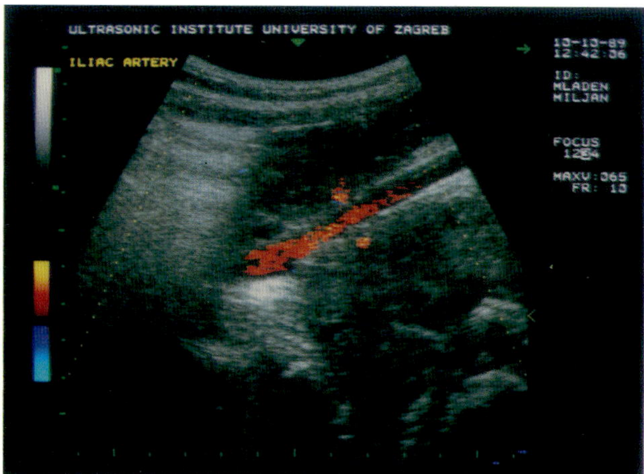

Figure 8.1 Color-coded blood flow (red) in the maternal iliac artery

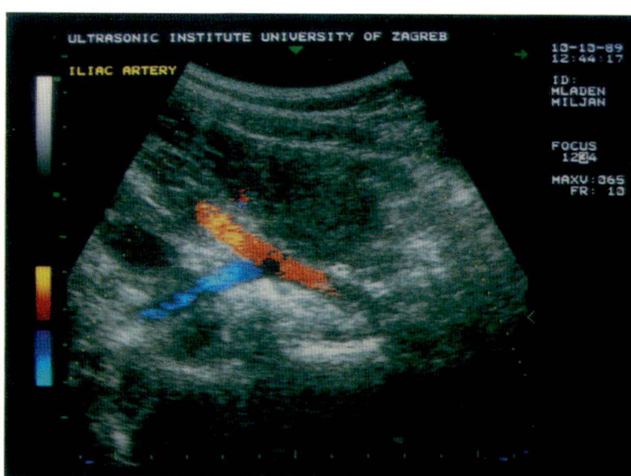

Figure 8.2 Color-coded blood flow in iliac (red) and uterine arteries (blue). Note that the uterine artery lies almost perpendicular to the iliac artery

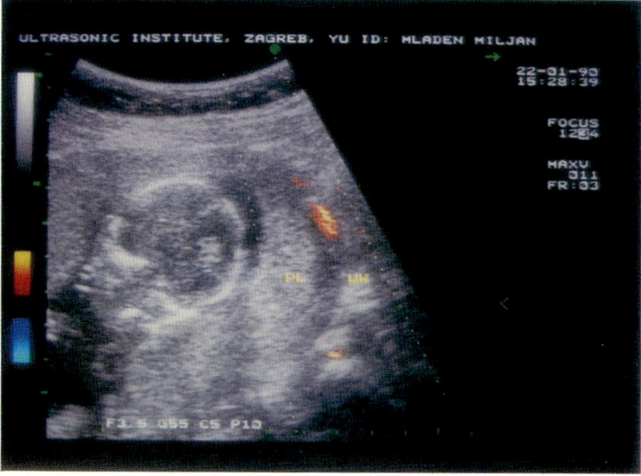

Figure 8.3 Color-coded blood flow in intervillous spaces (red). PL = placenta; UW = uterine wall

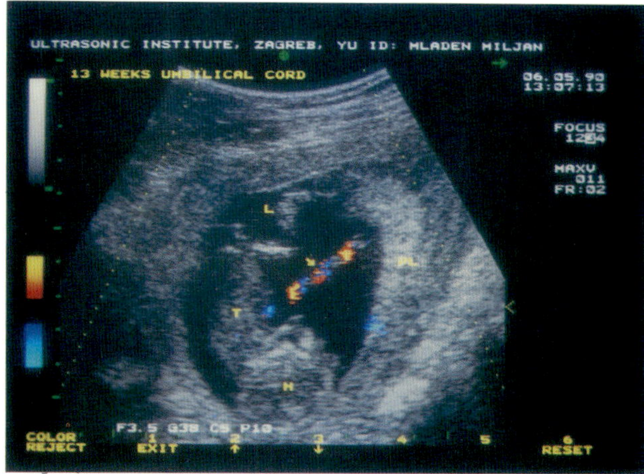

Figure 8.4 Blood flow in umbilical vessels in a 13-week pregnancy (arrow). H = fetal head; T = fetal trunk; L = fetal leg

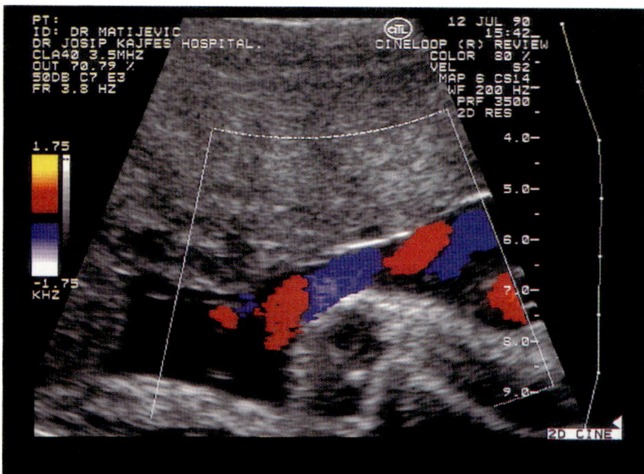

Figure 8.5 Umbilical cord blood flow visualized by color Doppler in a 28-week pregnancy

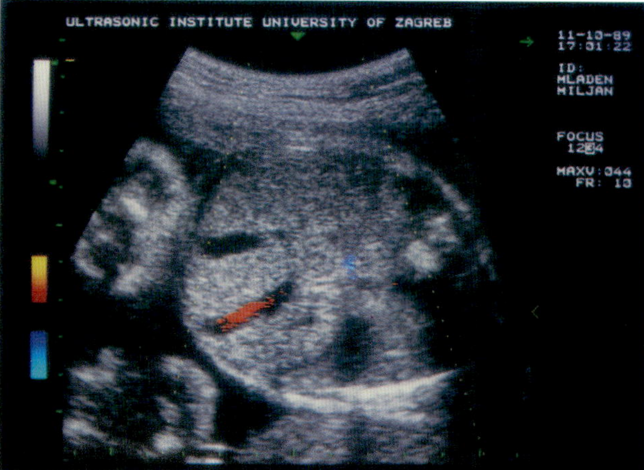

Figure 8.6 Blood flow in the intra-abdominal part of the umbilical vein (red)

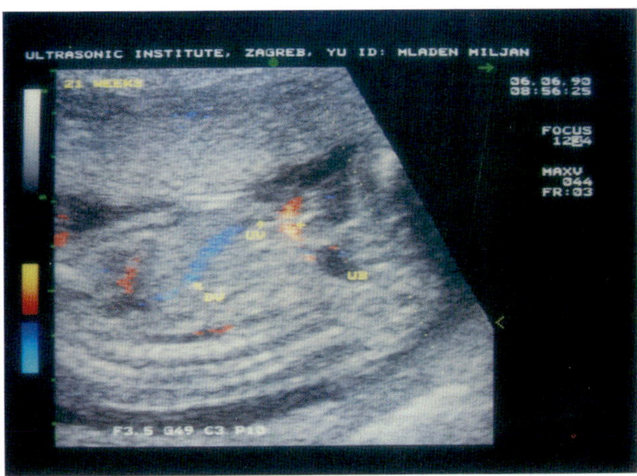

Figure 8.7 Color-coded blood flow in ductus venosus (DV) (blue). The red color represents blood flow in the umbilical artery. UV = umbilical vein; UB = urinary bladder

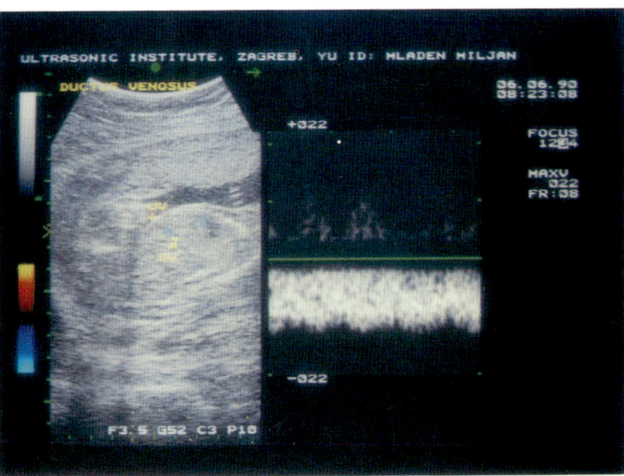

Figure 8.8 Two-dimensional color Doppler (left) and Doppler shift obtained from the ductus venosus (right) showing high end-diastolic flow

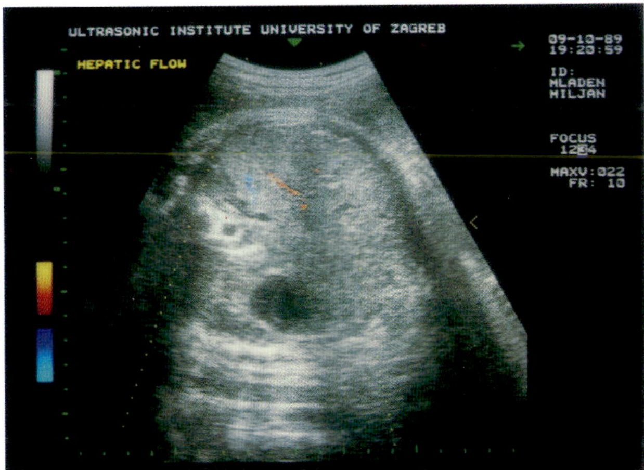

Figure 8.9 Color-coded blood flow through hepatic vessels (red and blue)

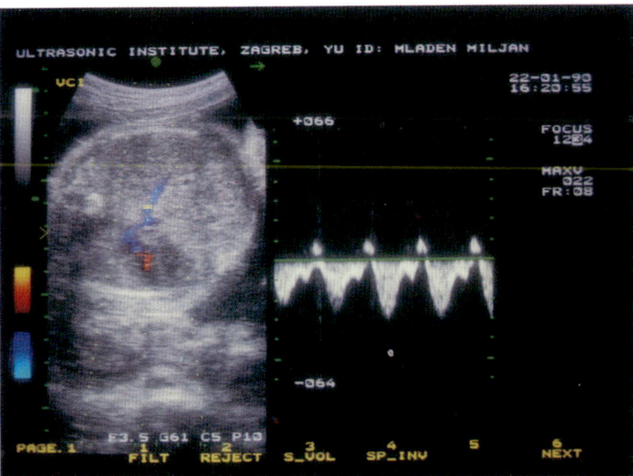

Figure 8.10 Color Doppler superimposed on the transverse scan of the fetal trunk (left), showing blood flow through the hepatic vein (blue). Doppler shift obtained from the main hepatic vein (right) showing typical biphasic flow

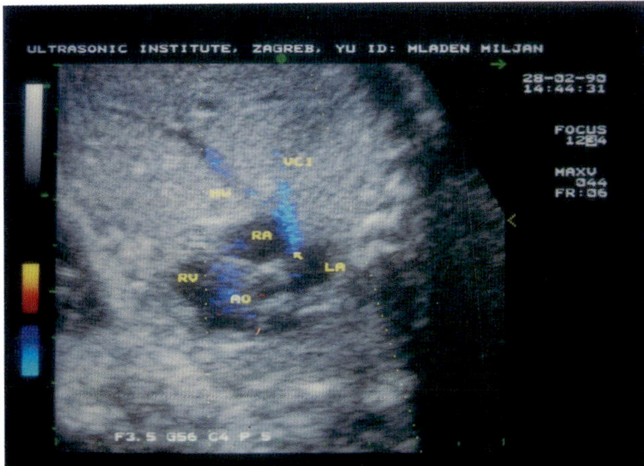

Figure 8.11 Color Doppler superimposed on the short axis of the fetal heart showing blood flow through the hepatic vein (HV) and inferior vena cava (VCI) (blue). The hepatic vein enters the VCI just before the former vein enters the right atrium (RA). LA = left atrium; RV = right ventricle; AO = aorta. Arrow indicates the position of foramen ovale

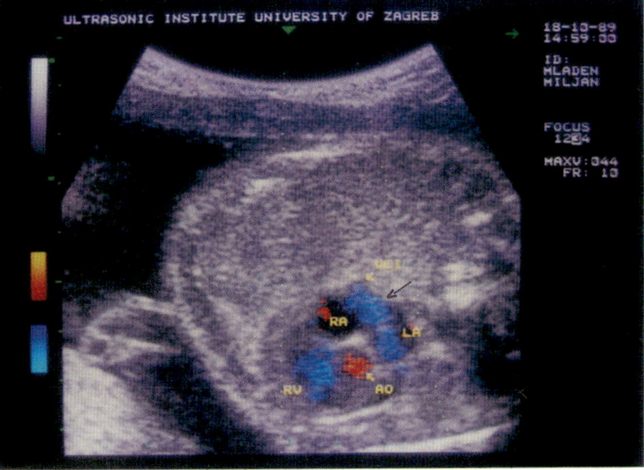

Figure 8.12 The short axis through the base of the heart demonstrating the bloodstream through foramen ovale from the right (RA) into the left atrium (LA) (blue, arrow). VCI = inferior vena cava; RV = right ventricle; AO = aorta

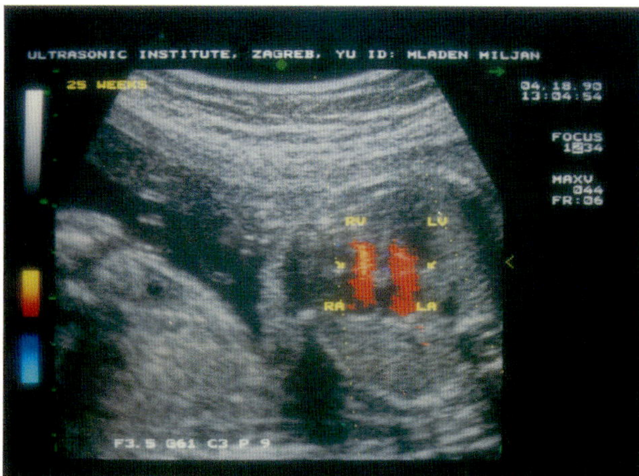

Figure 8.13 The four-chamber view of the fetal heart, demonstrating atrioventricular flows (red). RA = right atrium; RV = right ventricle; LA = left atrium; LV = left ventricle. The arrows indicate the level of atrioventricular valves

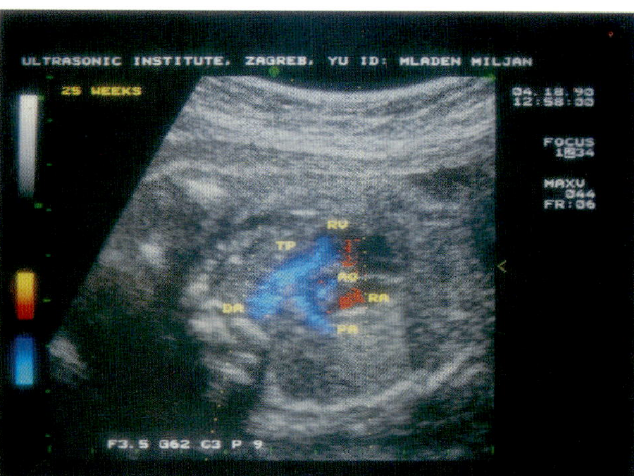

Figure 8.14 Short axis of the fetal heart at the level of the great arteries showing blood flow through the pulmonary trunk (TP), pulmonary artery (PA) and ductus arteriosus (DA) (blue). RV = right ventricle; AO = aorta; RA = right atrium

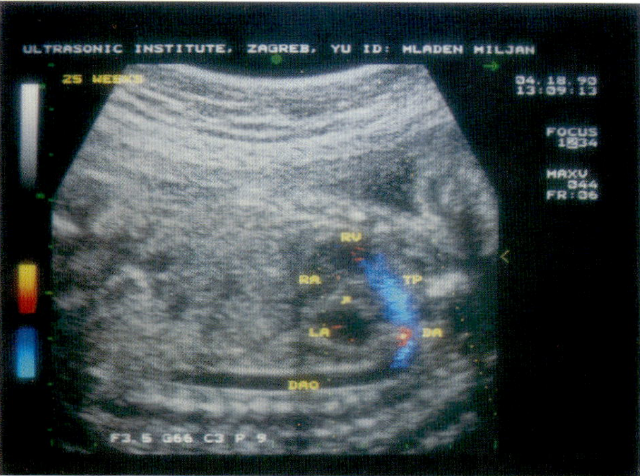

Figure 8.15 Color Doppler superimposed onto the ductus arteriosus plane demonstrating blood flow from the pulmonary trunk (TP), through the ductus arteriosus (DA), into the descending aorta (DAO) (blue). RV = right ventricle; RA = right atrium; LA = left atrium. The arrow indicates the proximal part of the ascending aorta

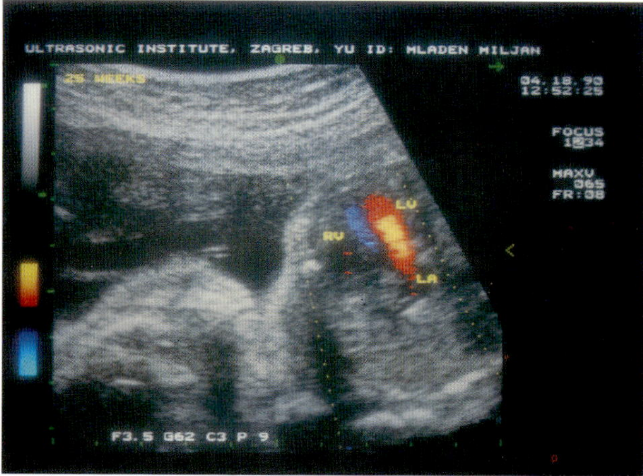

Figure 8.16 The bloodstream, originating from the pulmonary vein, passing through the left atrium (LA) and entering the left ventricle (LV) (red). RV = right ventricle

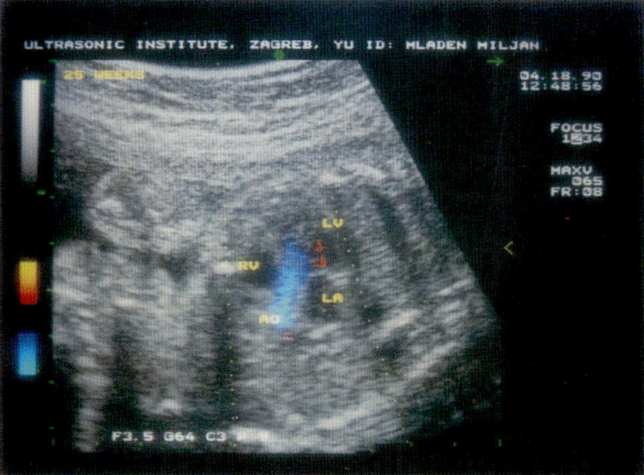

Figure 8.17 The long axis of the left ventricle (LV) demonstrating blood flow through the LV outflow tract and the proximal part of ascending aorta (AO) (blue). LA = left atrium; RV = right ventricle

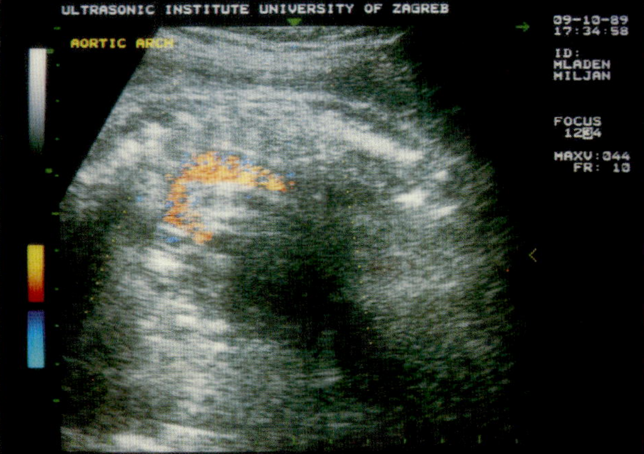

Figure 8.18 Parasagittal section through the fetal trunk (aortic arch plane) demonstrating blood flow through the aortic arch (red orange)

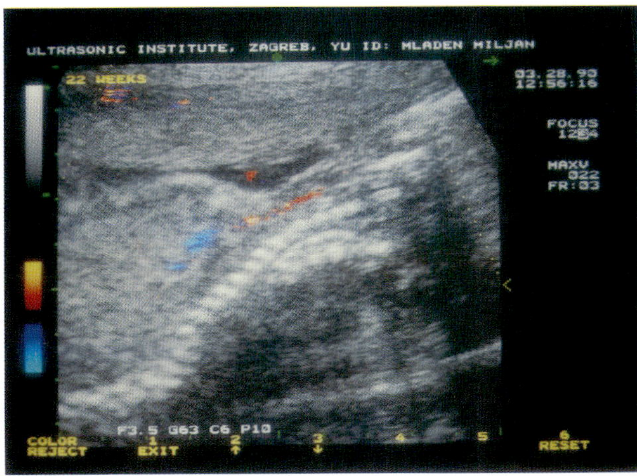

Figure 8.19 Parasagittal section through the fetal neck demonstrating blood flow in carotid artery (red)

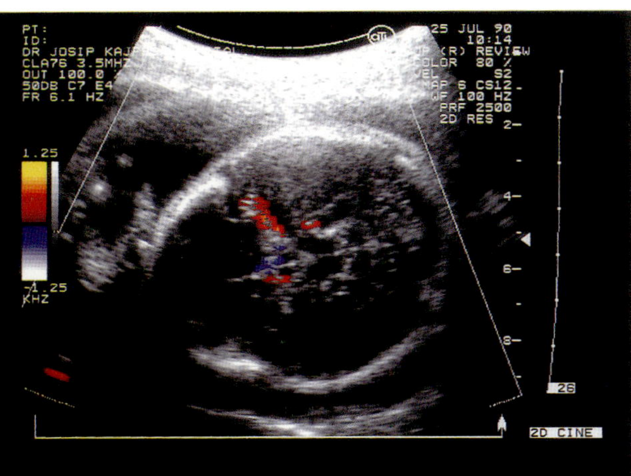

Figure 8.20 Color-coded blood flow in middle cerebral artery (red)

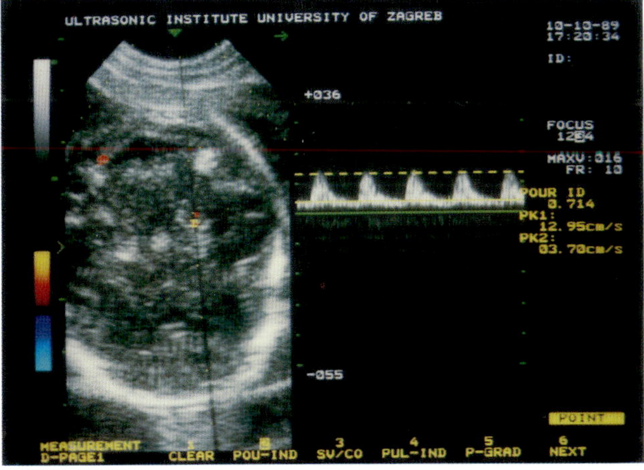

Figure 8.21 Two-dimensional color Doppler (left) and pulse Doppler signal (right) obtained from middle cerebral artery

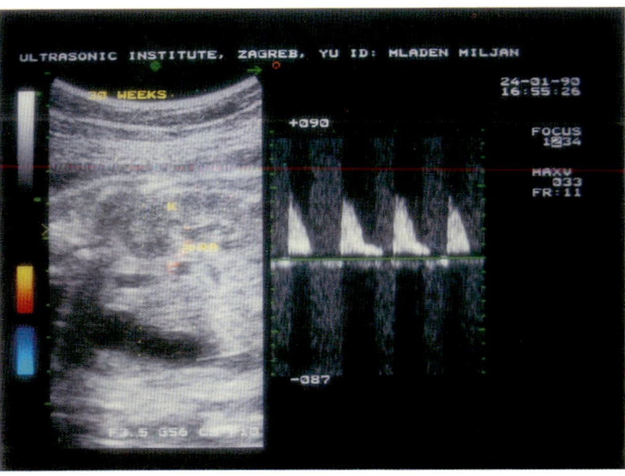

Figure 8.22 Color-coded (left, red color) and pulse Doppler (right) demonstrating blood flow in the fetal renal artery (RA). K = kidney

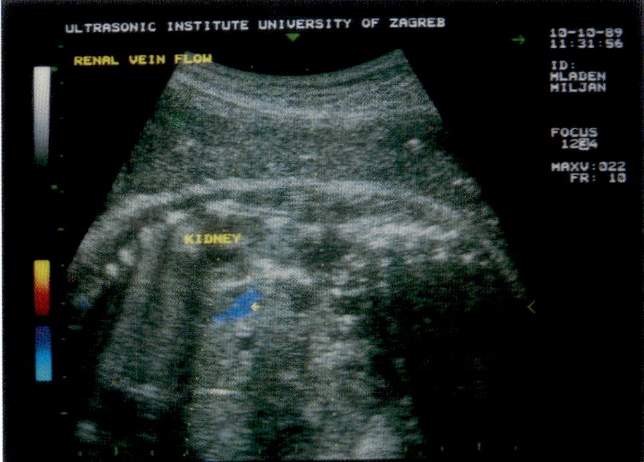

Figure 8.23 Color-coded blood flow in the fetal renal vein (arrow, blue)

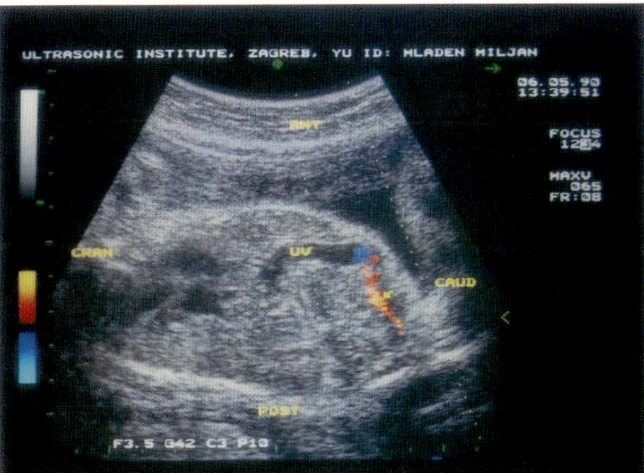

Figure 8.24 Blood flow in the umbilical artery detected by color Doppler (red, arrow). UV = umbilical vein; ANT = anterior; POST = posterior; CAUD = caudally; CRAN = cranially

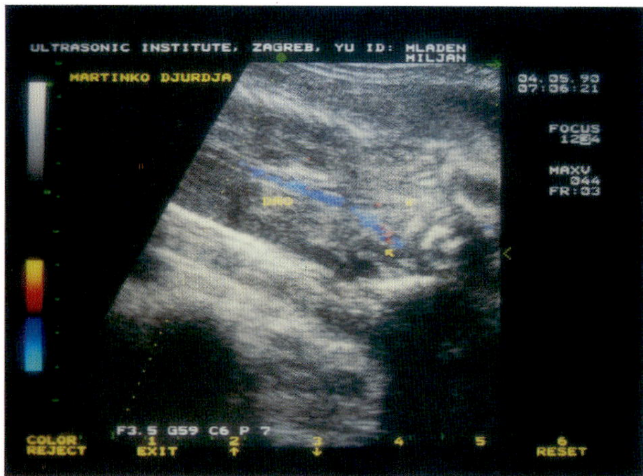

Figure 8.25 Blood flow in the descending aorta (DAO) and iliac arteries (blue). Iliac arteries are indicated by arrows

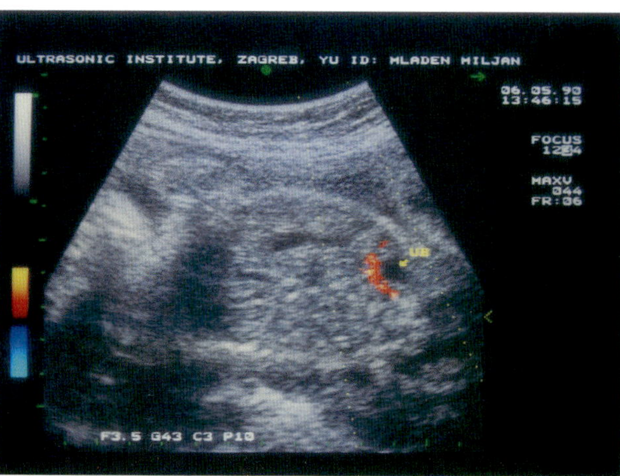

Figure 8.26 Perivesical blood flow detected by color Doppler (red). UB = urinary bladder

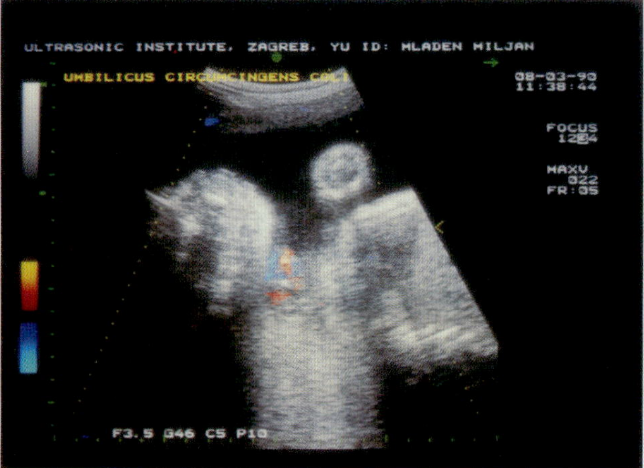

Figure 8.27 Blood flow through the umbilical vessels (red, blue) in a fetus with the umbilical cord being wrapped around its neck

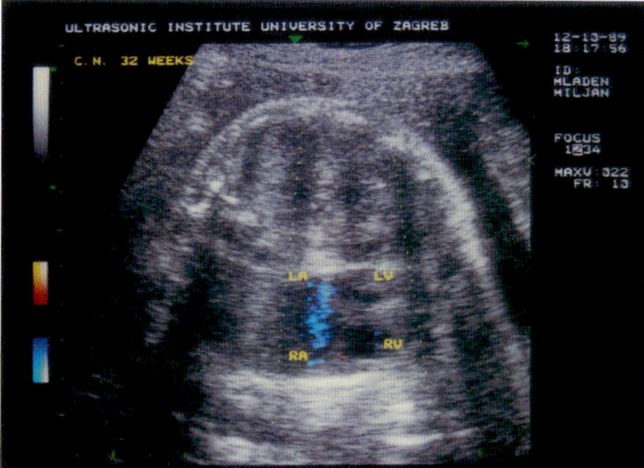

Figure 8.28 Blood flow from the left (LA) into the right atrium (RA) (blue) in a fetal heart with an atrial septal defect. LV = left ventricle; RV = right ventricle

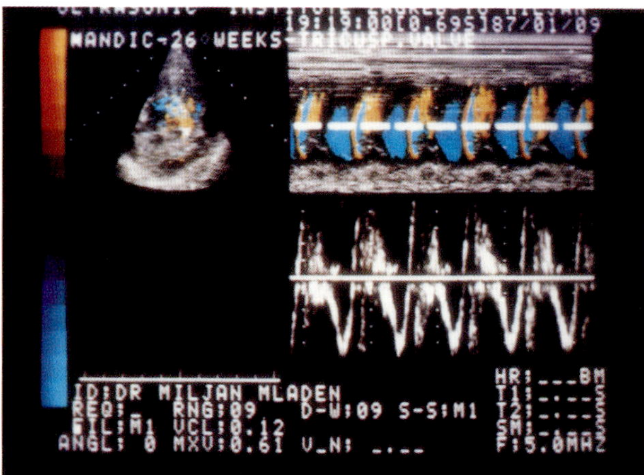

Figure 8.29 Two-dimensional color (left), M-mode color (upper right) and pulse Doppler (lower right) in a case of tricuspid insufficiency demonstrating blood regurgitation during ventricular systole (blue)

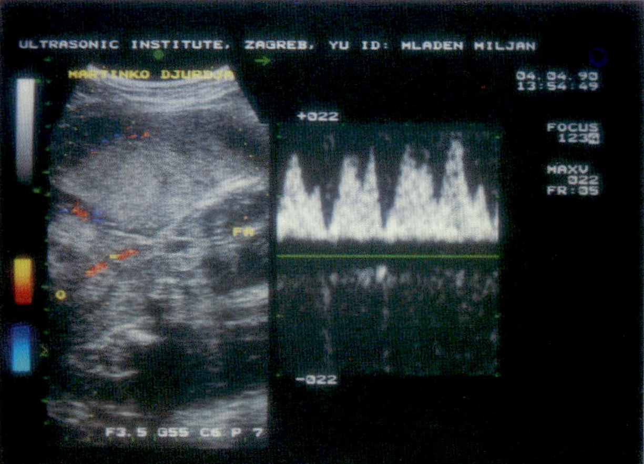

Figure 8.30 26-week-old fetus with an omphalocele. A blood vessel coming from the omphalocele and entering into the fetal abdomen was visualized by color Doppler (left, red). By placing sample volume gate inside the vessel, blood flow patterns typical of the hepatic vein are obtained by pulse Doppler (right), allowing precise identification of the blood vessel

portion, however, remains in the right atrium, and, after mixing with blood returning from the head and arms by way of the superior vena cava, passes through the tricuspid valve into the right ventricle (Figure 8.13). During diastole, blood is ejected from the right ventricle, through the pulmonary valve, into the pulmonary trunk (Figure 8.14). Since the resistance in the pulmonary vessels during fetal life is high, the main portion of this blood passes directly through the ductus arteriosus into the descending aorta (Figure 8.15), where it mixes with blood which is coming from the left ventricle. A smaller amount of blood from the pulmonary trunk enters the pulmonary arteries, and supplies the fetal lungs with oxygen (Figure 8.14).

Blood, which reaches the left atrium through the foramen ovale from the right atrium, mixes with a small amount of blood returning from the lungs through the pulmonary veins (Figure 8.16), and the blood enters the left ventricle through the mitral valve (Figure 8.13). During ventricular systole, blood is ejected from the left ventricle into the ascending aorta (Figure 8.17), and, passing through the aortic arch, enters the descending aorta (Figure 8.18).

Three major branches (truncus brachiocephalicus, left common carotid artery and left subclavian artery) originate from the aortic arch, supplying the fetal head and arms with blood. From the clinical point of view, blood flows through the common and internal carotid arteries (Figure 8.19) are of particular interest, because their measurement enables detection of circulatory redistribution, known as the 'brain-sparing effect', favoring brain perfusion in the presence of chronic fetal hypoxia. The same phenomenon may be detected by blood flow measurements in the three large intracranial vessels (anterior, medial and posterior cerebral arteries), which can be easily detected by color Doppler (Figures 8.20 and 8.21).

From the descending aorta, numerous branches originate, supplying in particular the fetal organs. Among them, the renal artery and vein (Figures 8.22 and 8.23) can be easily visualized. Of the blood in the descending aorta, 40–50% passes into the umbilical arteries (Figure 8.24), while the remainder circulates through the inferior half of the body. Among the vessels in the inferior part of the fetal abdomen, color Doppler allows visualization of blood flow in iliac arteries (Figure 8.25), as well as the flow in the blood vessels circuiting the fetal urinary bladder (Figure 8.26).

As well as the possibility that color Doppler offers in the evaluation of normal blood flow, it is tremendously important for the detection of various abnormal or pathological conditions, such as the umbilical cord wrapping round the fetal neck (Figure 8.27), pathological intracardiac shunts (Figure 8.28), or cardiac valvular diseases (Figure 8.29). It can also be used for elucidation of anomalous fetal circulation in extracardiac malformations, such as in the case of omphalocele (Figure 8.30), or for visualization of blood vessels overlapping the internal cervical os in cases of placenta previa (Figure 8.31).

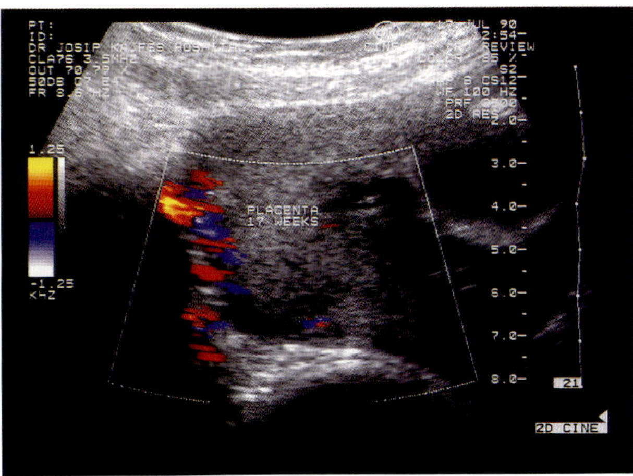

Figure 8.31 Color-coded blood flow through the vessels overlapping the internal cervical os in a case of placenta previa

REFERENCES

1. FitzGerald, D.E. and Drumm, J.E. (1977). Non-invasive measurement of fetal circulation using Doppler ultrasound: a new method. *Br. Med. J.*, **2**, 1450
2. Eldridge, M.W. (1981). Doppler ultrasound measurements of human fetal aortic blood flow. *Biomed. Sci. Instrum.*, **17**, 67
3. Jouppila, P. and Kirkinen, P. (1984). Increased vascular resistance in the descending aorta of the human fetus in hypoxia. *Br. J. Obstet. Gynaecol.*, **91**, 863
4. Maulik, D., Nanda, N.C. and Saini, V.D. (1984). Fetal Doppler echocardiography: methods and characterization of normal and abnormal hemodynamics. *Am. J. Cardiol.*, **53**, 572
5. Reed, K.L., Meijboom, E.J., Sahn, D.J., Scagnelli, S.A., Valdes-Cruz, L.M. and Shenker, L. (1986). Cardiac Doppler flow velocities in human fetuses. *Circulation*, **73**, 41
6. Allan, L.D., Chita, S.K., Al-Ghazali, W., Crawford, D. and Tynan, M. (1987). Doppler echocardiographic evaluation of the normal human fetal heart. *Br. Heart J.*, **57**, 528
7. Machado, M.V.L., Chita, S.C. and Allan, L.D. (1987). Acceleration time in the aorta and pulmonary artery measured by Doppler echocardiography in the midtrimester fetus. *Br. Heart J.*, **58**, 15
8. De Smedt, M.C.H., Visser, G.H.A. and Meijboom, E.J. (1987). Fetal cardiac output estimated by Doppler echocardiography during mid- and late gestation. *Am. J. Cardiol.*, **60**, 338
9. Reed, K.L., Anderson, C.F. and Shenker, L. (1987). Fetal pulmonary artery and aorta: two-dimensional Doppler echocardiography. *Obstet. Gynecol.*, **69**, 175
10. Wladimiroff, J.W., Tonge, H.M. and Stewart, P.A. (1986). Doppler ultrasound assessment of cerebral

blood flow in the human fetus. *Br. J. Obstet. Gynaecol.*, **93**, 471
11. Wladimiroff, J.W., Van Den Wijngaard, J.A.G.W., Degani, S., Noordom, M.J., Van Eyck, J. and Tonge, H.M. (1987). Cerebral and umbilical arterial blood flow velocity waveforms in normal and growth-retarded fetuses. *Obstet. Gynecol.*, **69**, 705
12. Woo, J.S.K., Liang, S.T., Lo, R.L.S. and Chan, F.Y. (1987). Middle cerebral artery Doppler flow velocity waveforms. *Obstet. Gynecol.*, **70**, 613
13. Veile, J-C. and Cohen, I. (1990). Middle cerebral artery blood flow in normal and growth-retarded fetuses. *Am. J. Obstet. Gynecol.*, **162**, 391
14. Van den Wijngaard, J.A.G.W., Groenenberg, I.A.L., Wladimiroff, J.W. and Hop, W.C.J. (1989). Cerebral Doppler ultrasound of the human fetus. *Br. J. Obstet. Gynaecol.*, **96**, 845
15. DeVore, G.R., Horenstein, J., Siassi, B. and Platt, L.D. (1987). Doppler color flow mapping: a new technique for the diagnosis of congenital heart defects. *Am. J. Obstet. Gynecol.*, **156**, 1054
16. Kurjak, A., Breyer, B., Jurković, D., Alfirevic, Z. and Miljan, M. (1987). Color flow mapping in obstetrics. *J. Perinat. Med.*, **15**, 271
17. Kurjak, A., Alfirevic, Z. and Miljan, M. (1988). Conventional and color Doppler in the assessment of fetal and maternal circulation. *Ultrasound Med. Biol.*, **14**, 337
18. Kurjak, A., Miljan, M., Jurković, D., Alfirevic, Z. and Žalud, I. (1989). Color Doppler in the assessment of the fetomaternal circulation. *Rech. Gynecol.*, **1**, 269

9 Ectopic Pregnancy

A. Kurjak and I. Žalud

Ectopic pregnancy still continues to be one of the major problems in gynecology. This is a pregnancy in which implantation occurs outside the endometrium and endometrial cavity, i.e. in the cervix, Fallopian tube, ovary, or abdominal or pelvic cavity. The true incidence is difficult to determine, but the reports indicate that during the last 15 years there has been a marked increase in ectopic gestation. In Sivin and Cooper's study of 35 496 American women, the rate of ectopic pregnancy almost doubled between 1965 and 1976[1]. Rubin *et al.* described an increase in rate from 4.5 to 9.4 per 1000 reported pregnancies[2]. In Sweden, the rate of ectopic pregnancy increased from 5.8 to 11.1 per 1000 conceptions in 15 years and in Great Britain, the rate per 1000 live births and therapeutic abortions increased from 3.2 to 4.3[3]. Despite the introduction of modern diagnostic methods such as ultrasonography and rapid and specific β-hCG assays which lead to early surgical intervention, ectopic gestation is still the major cause of maternal mortality in the first trimester of pregnancy[4]. The death rate is approximately 1 in 1000 cases, delay in making the diagnosis being the major underlying reason for this[5]. Ectopic pregnancy accounts for about 26% of all maternal deaths in the United States, a fetal wastage of nearly 100%, and a high incidence of maternal morbidity[6].

The basic reason for the increase in the rate of ectopic pregnancy is the fact that more adolescent girls are sexually active; with this comes sexually transmitted disease as well as pregnancy. The micro-organisms causing ectopic pregnancy are usually the *Chlamydiae* and *Neisseria gonorrhoeae*, but other microbes may be implicated in the resulting salpingitis as well[7]. Therefore, it becomes increasingly important to consider ectopic pregnancy in the differential diagnosis of a teenager presenting with abdominal pain and irregular vaginal bleeding. The chance of recurrence varies from 5 to 20%[7]. In general, the likelihood of ectopic pregnancy increases with previous tubal disease, ectopic pregnancy or induced abortion. Intrauterine devices do not prevent ectopic pregnancy. Some women with a previous ectopic pregnancy develop problems of infertility and never deliver a living child. According to Curran's projection, by the year 2000 at least 10% of all females of reproductive age will become involuntarily sterile as a result of the sequelae of pelvic inflammatory disease; more than 3% will experience an ectopic gestation[8]. Ongoing improvements and results of *in vitro* fertilization programs should not diminish efforts to improve prevention, diagnosis and treatment of ectopic pregnancy.

PATHOGENESIS

At least 97% of ectopic gestations occur in the Fallopian tube, with the majority sited in the ampulla or isthmus. Implantation in the interstitial portion of the tube is rare but extremely dangerous. Less common ectopic sites are in the cervix, in a rudimentary horn of a bicornuate uterus, in the ovary, in the broad ligament, or on the peritoneum. Cases of the latter may be primary implantations but are more commonly believed to be secondarily implanted following rupture of a tubal pregnancy. Occasionally bilateral tubal pregnancies occur, and ectopic pregnancy may also coexist with an intra-uterine pregnancy[9]. Implantation of the conceptus in the ectopic position may occur because its passage to the uterus is retarded or because factors in the tube or in the conceptus facilitate premature implantation.

In ectopic tubal pregnancy, the fertilized ovum implants beneath the epithelium of the oviduct to form a fluid-filled gestational sac lined with trophoblastic tissue in the wall of the tube. Since the oviduct has only a thin layer of muscle, the trophoblastic cells that burrow deep into the tubal epithelium distend the oviduct and eventually cause rupture. The gestational sac within the tube of a ruptured ectopic pregnancy is usually surrounded by fluid or blood due to erosion of adjacent vessels. In the vast majority of cases, the separation of the decidua from the wall of the oviduct causes death of the fetus. In rare cases, the fetus may survive an attempt at abortion by reimplantation within the abdomen and re-establishment of the blood supply from the omentum or the mesentery.

Mild uterine enlargement and endometrial

hypertrophy are usually present with an ectopic pregnancy and can occasionally be detected clinically. If dilatation and curettage is performed on a patient with an ectopic pregnancy, only decidua without chorionic villi will be obtained.

TRANSABDOMINAL ULTRASOUND

Early diagnosis of any ectopic pregnancy still remains a challenge. Although diagnostic ultrasound has played an increasing role in the assessment of patients suspected of having ectopic pregnancy, its major role has been in excluding an intrauterine gestation in patients confirmed to be pregnant.

The sonographic findings that may by encountered in patients with ectopic pregnancy are divided into those that are considered 'diagnostic' and those thought to be 'suggestive'[10]. The 'diagnostic' signs are absence of an intrauterine gestational sac bordered by two layers of decidua, extrauterine and extraovarian adnexal mass, and fetal heart beats and motions (Figures 9.1 and 9.2). The 'suggestive' features of ectopic pregnancy are an 'enlarged' uterus with thick, echogenic endometrium in the case of unruptured pregnancy, and blood or organized clot in the cul-de-sac, or pericolic recesses in the case of ruptured pregnancy. The transabdominal sonographic diagnosis of an ectopic pregnancy can be made with greater confidence if more than one sonographic sign is present.

Because the size of the ectopic pregnancy can be small (less than 1 cm in size), its sonographic detection as an adnexal mass is variable. In most patients with an unruptured ectopic pregnancy, an adnexal mass separate from the ovary, which has a small anechoic center, can be identified[10]. Rarely, fetal heart motion can be detected within the ectopic gestational sac which arises from a live fetus. In addition, the uterus in a patient with an ectopic pregnancy typically has an echoic endometrium, a sign of a decidual thickening associated with ectopic pregnancy. Such a finding is described as a 'pseudogestational sac'[11]. In contrast to the two concentric rings of decidua that can be identified in an early intrauterine pregnancy, the decidual cast of an ectopic pregnancy has only one layer of decidua. A double decidual sac can occasionally be mimicked by separation of the decidua from the myometrium prior to expulsion of a decidual cast associated with an ectopic pregnancy. A double decidual sac can also be encountered in patients with an incomplete abortion. Furthermore, approximately 6% of the patients with a single decidual layer have a viable intrauterine pregnancy. Obviously, a 'pseudogestational sac' may be the source of many failed diagnoses, both positive and negative. The sonographic detection of the double decidual layer is highly dependent upon the resolution of the scanner used and the scanning ability and experience of the individuals who perform the examination.

As well as meticulous sonographic evaluation of the uterus and adnexa, one should carefully evaluate the cul-de-sac for the presence of intraperitoneal blood in patients suspected of having an ectopic pregnancy. The presence of fluid within the cul-de-sac most frequently indicates the presence of a ruptured ectopic pregnancy. The sonographic appearance of intraperitoneal blood varies from anechoic to hyperechoic, probably depending upon the amount of organization that has occurred within the clot. Unclotted blood is typically anechoic and becomes more echogenic as it organizes.

The adnexal mass arising from an ectopic pregnancy can be associated with a corpus luteum. A corpus luteum might be differentiated from an adnexal mass resulting from an ectopic pregnancy by its location within the ovary and more rounded configuration[10]. However, corpus luteum can also be a source of false diagnosis of ectopic pregnancy.

Transabdominal ultrasound as a specific diagnostic tool for ectopic pregnancy is good enough only in typical cases. However, many ectopic pregnancies do not present gestational sacs and live embryos (Figures 9.3 and 9.4). Whereas Gleicher *et al.* in a small study were able to see the gestational sac in the adnexa in eight of nine patients with an ectopic pregnancy[12], other authors reporting larger series have been much less successful. Mahony *et al.* described abnormal adnexal findings in 80% of patients with an ectopic pregnancy, but many of these were non-specific and commonly present in other conditions or indeed in normal patients, e.g. free intraperitoneal fluid[13]. They were able to see the ectopic gestational sac in only five of 35 patients who subsequently proved to have an ectopic pregnancy.

When ultrasound findings are correlated with hormone tests, valuable clinical information is obtained in assessing women suspected of having an ectopic gestation[14]. However, because of the difficulty in determining the appropriate timing of surgery in patients with a positive pregnancy test and empty uterus on ultrasound, some authors have advocated the use of a 'discriminatory zone'[15, 16]. Unfortunately, only some patients have serum β-hCG levels above this range when they first present[15]. If the discriminatory zone is to be used, each laboratory must develop its own range. In addition, patients with a recent complete abortion will continue to present a diagnostic dilemma as even the urinary hCG may remain positive for up to 3 weeks after termination of pregnancy[17].

TRANSVAGINAL ULTRASOUND

Despite the contribution made by transabdominal diagnostic ultrasound in the management of patients suspected of having an ectopic pregnancy, far greater assistance could be provided to the clinician if the ectopic pregnancy was able to be demonstrated in the adnexa with more reliability. High-frequency transvaginal probes with their better resolution generate higher quality images than traditional abdominal probes. The improved images obtained involve all the target organs and spaces which are scanned in the work-up of ectopic pregnancy[18].

The detection of an intrauterine gestational sac can be achieved by the transvaginal approach earlier (usually as early as 16 days after conception) and easier than by the transabdominal route[19]. When using a transabdominal probe, one may find it difficult to differentiate a 'pseudogestational sac' from a true gestational sac. Using transvaginal sonography, these echoes were found to originate from a local blood clot or from a central sonolucent area outlined by a thick endometrium. If an intrauterine gestation cannot be identified, one should proceed to scan the Fallopian tubes.

The Fallopian tube may be identified if it is outlined by fluid or if its lumen contains a fluid phase[20]. The diagnosis of tubal pregnancy by transvaginal ultrasound can be based on three different characteristic patterns[19]. The first depends on the capability of transvaginal ultrasound to recognize gestational structures in the Fallopian tube using the same criteria as in an intrauterine pregnancy. Therefore, one may rely upon the detection of structures such as a gestational sac, a yolk sac or an embryo within the Fallopian tube when making the diagnosis. The second pattern is based on the recognition of a Fallopian tube with an amorphous content. At surgery this content has usually been found to be composed of blood flow, blood clots and gestational material. The third pattern is based on indirect signs such as an empty uterus, a positive serum β-hCG test, and detection of blood and/or blood clots in the cul-de-sac. Rottem *et al.* have found the first pattern in nearly 50% of investigated suspected ectopic pregnancies, the second pattern in nearly 40% and the third pattern in about 10% of patients[19]. Fetal heart beats were observed in nearly 23% of cases.

DeCrespigny also demonstrated the value of transvaginal sonography in the diagnosis of ectopic pregnancy[21]. Such a diagnostic approach allowed earlier demonstration of an intrauterine gestation without confusion with a 'pseudogestational sac'. Of 36 patients with an ectopic pregnancy, this was strongly suspected in 29 (81%). A live extrauterine embryo was seen in eight (22%), an ectopic trophoblast or a gestation-sac-like structure in 19 (53%), and a pelvic hematoma in two (6%). Other authors also advocated transvaginal sonography as the technique of choice in patients suspected of having an ectopic pregnancy[22–31].

Even though transvaginal sonographic examination has become the most reliable instrument in the diagnosis of normal and abnormal intrauterine pregnancy, unfortunately it has not yet reached such a level of success in the diagnosis of ectopic pregnancy. However, demonstration of an ectopic pregnancy with transvaginal ultrasound allows a more precise preoperative diagnosis, so it is likely to allow earlier surgical intervention with resulting minimization of tubal damage. However, many ectopic pregnancies do not develop an embryo so that these specific findings are not seen and the presence of an adnexal mass may be ambiguous. A prudent approach to the management of ectopic pregnancy should take into consideration a number of limitations of transvaginal sonography in the work-up of this condition. The main limitation is the risk of obtaining a false positive diagnosis of the corpus luteum. A false-negative diagnosis may be made in the case of an ectopic gestation that is located outside the true pelvis or in the event of a very small tubal gestation. Thus, other means to identify the nature of an adnexal mass become important and these are still controversial.

DOPPLER ULTRASOUND

The first transabdominal Doppler study of ectopic pregnancy was recently reported[32]. A viable ectopic embryo was documented in only 14% of cases. High-velocity flow, which suggested the presence of an ectopic pregnancy, was documented in 54% of patients. The positive predictive values were 47% for B-mode imaging alone and 85% for Doppler. The negative predictive values were 60% for imaging alone and 81% for Doppler. The limitations of transabdominal Doppler imaging in the diagnosis of ectopic pregnancy currently include the need for a full-bladder technique and the necessity for considerable operator experience in locating the placental flow and optimizing the signal. The mentioned authors concluded that these difficulties and the false-positive rate could be reduced by the addition of color and pulsed Doppler imaging to the transvaginal probe. This would combine the advantages of transvaginal sonography with the tissue characterizing ability of Doppler imaging (Figure 9.5). Such sensitive equipment has recently been produced and transvaginal color Doppler has now been tested for the assessment of ectopic pregnancy[33–41].

Transvaginal color Doppler can help to

characterize the nature of the adnexal mass, thus permitting preoperative diagnosis when the ectopic embryo and its characteristic heart beat cannot be visualized (Figures 9.6–9.13). We have studied the value of transvaginal color Doppler in the blood flow detection of ectopic pregnancy (Table 9.1)[38]. Ectopic pregnancy was defined as ectopic color flow, usually very prominent and randomly dispersed inside the solid part of the adnexal mass and clearly separated from ovarian tissue and corpus luteum. Pulsed Doppler waveform analysis showed a very low-impedance signal and the calculated resistance index was below 0.40 due to the increased end-diastolic flow. Taylor *et al.* hypothesized that such low-impedance flow around an ectopic gestation results from the hemodynamics of early placentation[42]. The trophoblast actively invades the maternal tissue, eroding blood vessels which bleed into the intervillous space and thereby start a primitive trophoblastic circulation. The hemodynamics involved in this provide the basis for the observed characteristic flow. The fact that the intervillous space lacks a muscle layer could also explain the low-resistance blood flow. The brightness of color is usually high, indicating a high velocity of ectopic flow.

There is great similarity between the characteristics of blood flow in the case of ectopic pregnancy and the abnormal Doppler signals seen in the neovascularity of tumors. In tumors, there is histological evidence to suggest that relatively high velocities result from shunting across large pressure gradients, the only possible source of such kinetic energy[43]. Very low impedance, giving small systolic–diastolic modulation, implies flow in vessels that are lacking muscular support. In Taylor's study, histological examination of the products of conception showed similar vascular morphology, with large tissue spaces distended with blood[32]. Lacunar spaces, the precursors of the intervillous space, are present at 10 days after conception. By the 3rd week, the circulation between the placenta and the embryo is probably functional. By 6 weeks, some maternal spiral arteries open directly into the intervillous space. A pressure gradient could generate the high velocities. Since most of the peripheral resistance resides in the arterial muscle, the low impedance of the ectopic blood flow signal is likely to result from the lack of muscle in the walls of the intervillous space.

Basically, the results obtained by transvaginal color Doppler diagnosis of ectopic pregnancy have been good enough to encourage its clinical application (Table 9.1). We have examined 148 patients with suspected ectopic pregnancy. Nine false-negative findings were observed in cases of tubal abortion. Color flow could not be visualized, probably because of an inactive trophoblast. In these patients the serum β-hCG levels have been measured and have been found to be decreased. The highest value was 720 mIU/ml and the lowest 80 mIU/ml. Two false-positive findings were obtained in the case of a corpus luteum cyst. We suggest that the current policy should be to delay surgical management if there is no ectopic blood flow outside the empty uterus in amenorrheic patients. On the contrary, if there is color flow in the adnexal region with a resistance index <0.40, the patient should be scheduled for laparoscopy regardless of the clinical signs. The hypothesis is that the absence of color flow from the ectopic pregnancy and corpus luteum may indicate that the ectopic pregnancy is no longer viable. There is no doubt that some ectopic embryos die and are resorbed. Color Doppler signals might be helpful in predicting which ectopic embryos could be treated expectantly. Nevertheless, in the absence of free fluid in the cul-de-sac, this policy suggests that the population without color flow involves ectopic pregnancies which are non-viable, and, therefore, best suited for medical therapy, i.e. methotrexate[42].

Table 9.1 Diagnostic value of transvaginal color Doppler in the detection of ectopic pregnancy ($n = 148$)

Color flow and RI < 0.40	*Ectopic pregnancy*		
	Yes	*No*	*Total*
Yes	64	2	66
No	9	73	75
Total	73	75	148

Positive predictive value = 97%; negative predictive value = 89%; sensitivity = 88%; specificity = 97%; accuracy = 93%

The β-hCG levels were correlated with the resistance index of ectopic blood flow and the last menstrual period[38]. However, statistical analysis indicated that this relationship cannot be used effectively as diagnostic information. It does suggest that the resistance in the uteroplacental vessels is decreasing when there is progressive trophoblastic growth. There were nine cases of ectopic pregnancy in which there was no color flow demonstrated. In all these women the β-hCG was less than 1000 mIU/ml. The operative impression was that of probable tubal abortion. This diagnosis was usually made when there was an empty Fallopian tube, clots, blood in the cul-de-sac, and no visible embryo. These findings may be useful in the formulation of a plan of management.

CORPUS LUTEUM IN ECTOPIC PREGNANCY

Angiogenesis in corpus luteum occurs in physiological circumstances in each menstrual cycle.

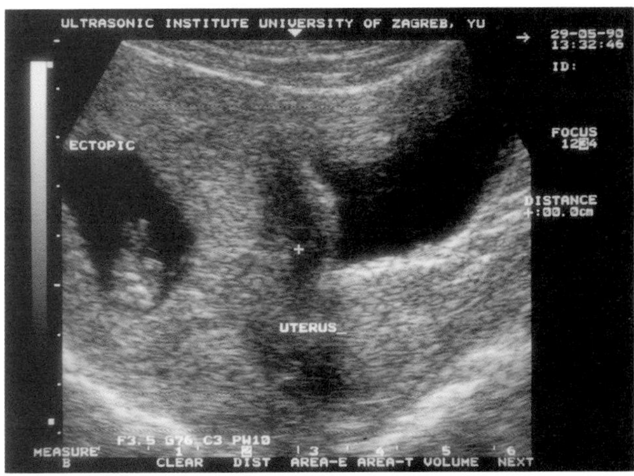

Figure 9.1 A transabdominal scan. An empty uterus, ectopic gestational sac, embryonal echo and embryonal heart beats (on real-time scan) are clearly visible. This is a typical finding of an ectopic pregnancy

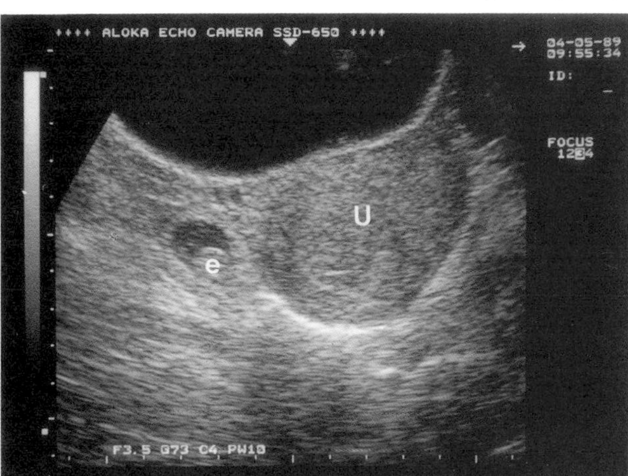

Figure 9.2 Another example of a typical ultrasound finding of an ectopic pregnancy. Empty uterus and ectopic gestation are easily visualized. e = ectopic embryo; U = empty uterus

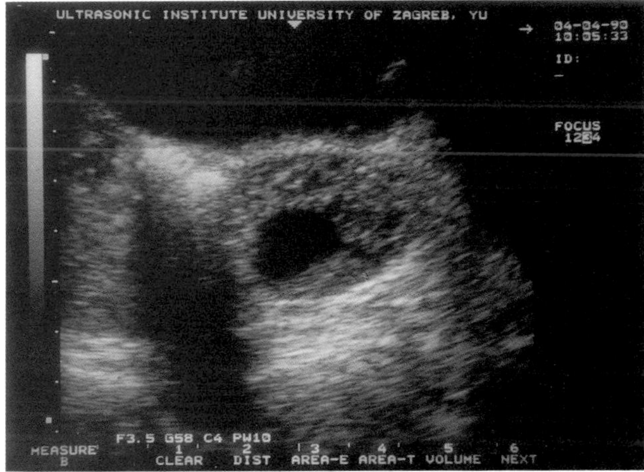

Figure 9.3 The left ovary visualized by transabdominal scan. Small transonic structure represents a corpus luteum cyst and should not be mistaken for the ectopic gestational sac

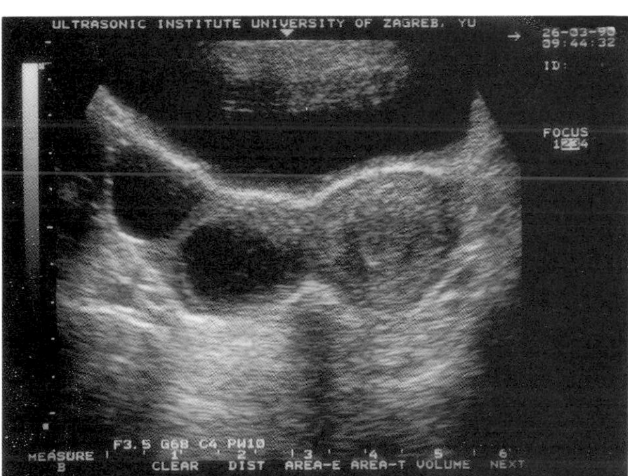

Figure 9.4 An empty uterus and a bilocular cystic tumor in the right adnexa, another non-specific ultrasound feature in suspected ectopic pregnancy

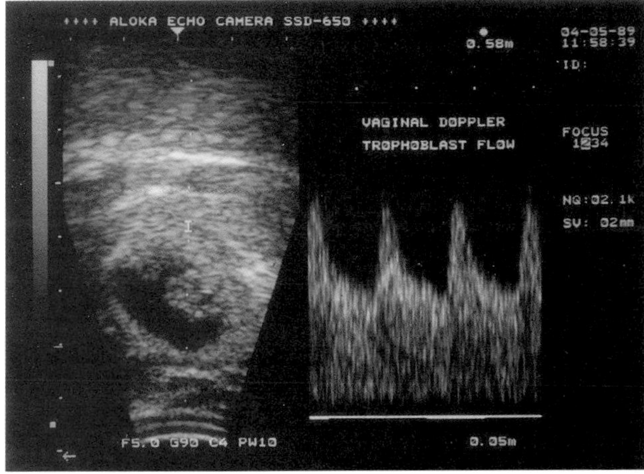

Figure 9.5 Transvaginal sonogram presenting ectopic gestation and a live embryo. Ectopic blood flow was detected by pulsed Doppler (right)

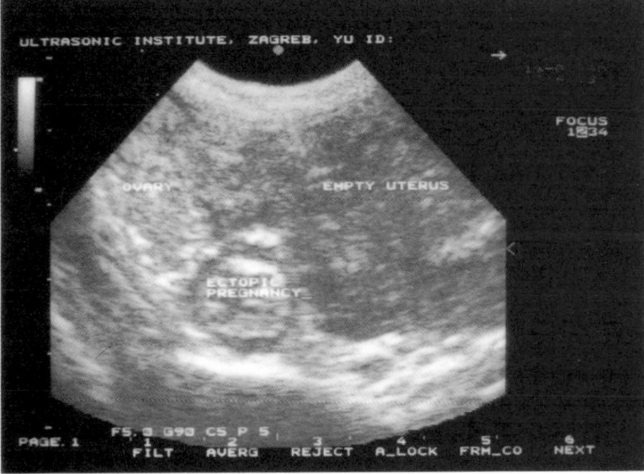

Figure 9.6 A transvaginal scan of an empty uterus, the right ovary and a solid adnexal mass suspected to be an ectopic gestation. This is a non-specific ultrasound finding

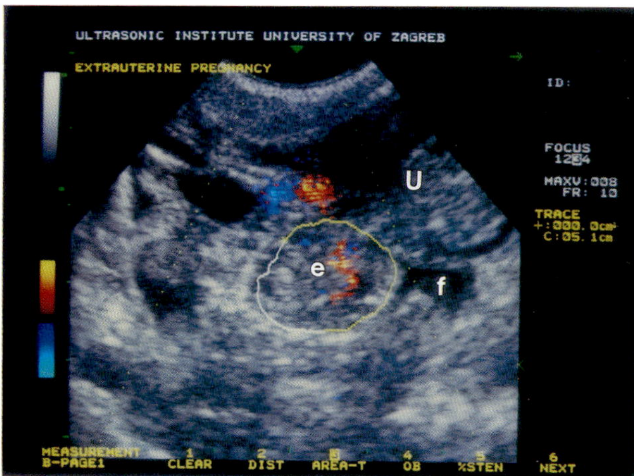

Figure 9.7 The same patient. A highly vascularized adnexal mass suspected to be an ectopic pregnancy. Transvaginal color Doppler helps to diagnose the ectopic pregnancy in non-specific ultrasound findings. U = empty uterus; e = ectopic pregnancy; f = free fluid

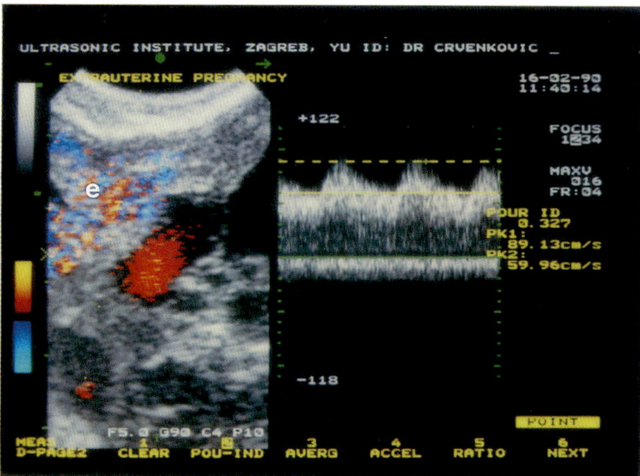

Figure 9.8 Another example of suspected ectopic pregnancy represented by a solid adnexal mass. Transvaginal color Doppler shows ectopic blood flow. Pulsed Doppler (right) shows high velocity and very low resistance of flow (RI = 0.327). These are typical Doppler findings of ectopic pregnancy. The diagnosis was confirmed by surgery. e = ectopic gestation

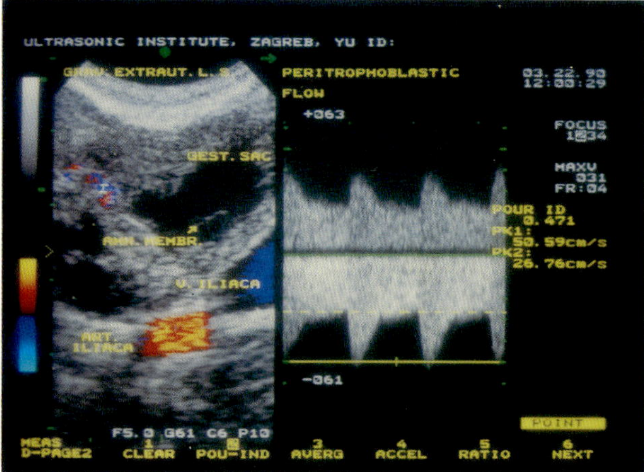

Figure 9.9 Transvaginal color Doppler findings of ectopic pregnancy

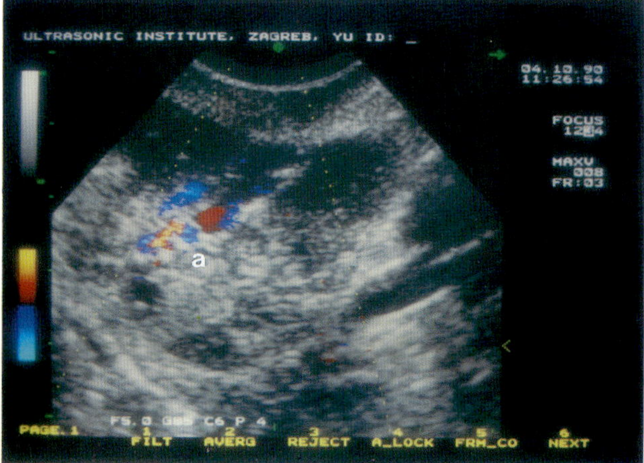

Figure 9.10 An adnexal mass suspected to be an ectopic pregnancy. Transvaginal color Doppler shows ectopic blood flow. a = adnexal mass

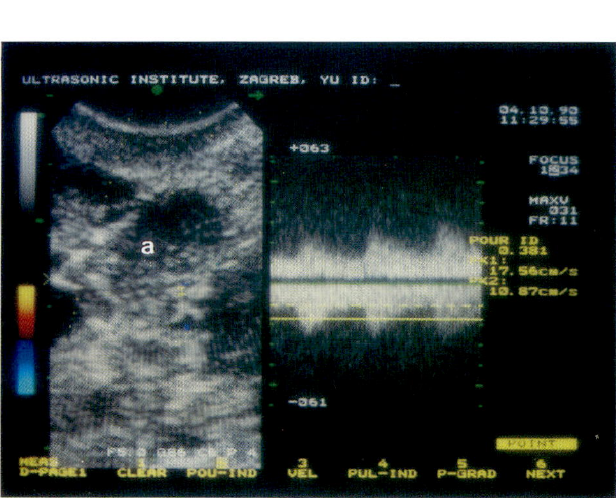

Figure 9.11 The same patient. Pulsed Doppler (right) analysis shows very low-resistance flow. Doppler findings help to solve the dilemma: is it an ectopic pregnancy or another adnexal tumor? a = adnexal mass

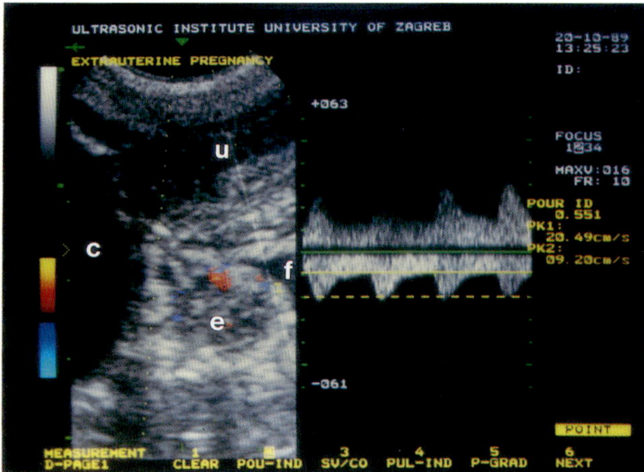

Figure 9.12 Color and pulsed Doppler characterization of a suspected ectopic pregnancy (e). u = empty uterus; c = corpus luteum cyst; f = free fluid

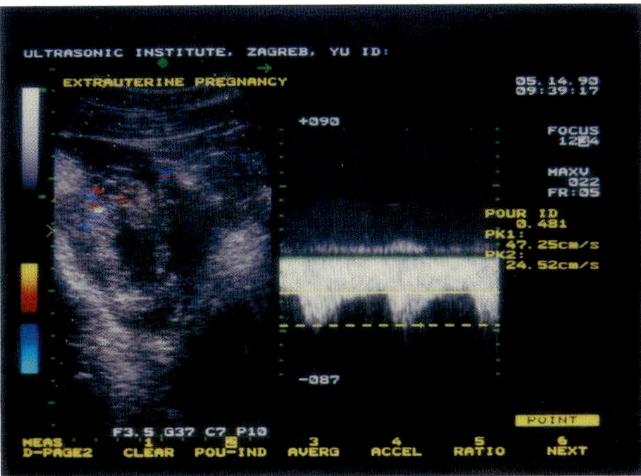

Figure 9.13 Another example of an ectopic pregnancy

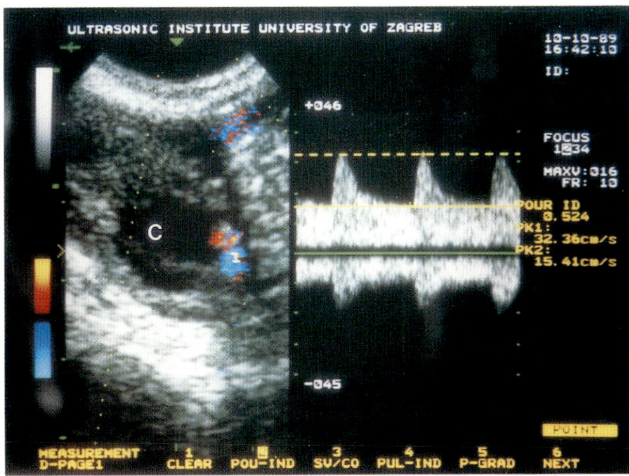

Figure 9.14 Blood flow was detected at the periphery of the small ovarian cyst. Waveform analysis (right) shows moderate velocity and low resistance flow (RI = 0.524). This is a typical Doppler finding of corpus luteum neovascularization, not to be mistaken for an ectopic gestation. c = corpus luteum cyst

However, a certain overlap between luteal blood flow and color flow in the case of ectopic pregnancy or ovarian malignancy could be expected (Figure 9.14)[32, 41, 43, 44]. Using high quality B-mode transvaginal sonography and superimposed color Doppler flow, it should be possible to distinguish between luteal and ectopic pregnancy blood flow. Only ovarian pregnancy can produce a diagnostic challenge. However, our results show that corpus luteum flow is significantly higher than ectopic blood flow and can be used as a diagnostic parameter (Table 9.2).

It was suggested that absence of luteal flow is inconsistent with a viable normotopic as well as ectopic pregnancy[32]. Since an increasing resistance index reflects some reduction in the blood flow distal to the point at which the Doppler recording was made, namely in the corpus luteum, one might postulate that this reflects a reduction in corpus luteum function which is incompatible with pregnancy. Transvaginal color Doppler seems to be sufficient enough to detect luteal flow easily and accurately in non-pregnant women and in normal and abnormal early pregnancy (Table 9.3). Furthermore, preliminary results of resistance index analysis of such color flow show sufficient difference to distinguish between luteal flow in non-gravid, pregnant and ectopic pregnant patients (Table 9.4). It is justifiable to expect that further investigations will confirm transvaginal color imaging as a diagnostic tool for evaluation of normal morphological and functional features of the ovary in ectopic pregnancy. Corpus luteum blood flow analysis could play an important role in such an approach.

Table 9.2 The resistance index of corpus luteum blood flow and ectopic pregnancy

	n	*Resistance index*	*2 SD*
Corpus luteum	59	0.48	0.07
Ectopic pregnancy	64	0.36	0.04

$t = 11.5; p < 0.01$

Table 9.3 Detection rate of corpus luteum blood flow in non-pregnant and pregnant women ($n = 243$)

		Corpus luteum blood flow	
Patients	*n*	*Detected*	*Not detected*
Non-gravid	70	65 (92.9%)	5 (7.1%)
Early pregnancy	100	83 (83.0%)	17 (17.0%)
Ectopic	73	59 (80.8%)	14 (19.2%)

Table 9.4 Resistance index (RI) in corpus luteum blood flow ($n = 207$)

Patients	*n*	*RI*	*2 SD*
Non-gravid	65	0.42^a	0.12
Early pregnancy	83	0.53^b	0.09
Ectopic	59	0.48^c	0.07

$t_{ab} = 6.16, p < 0.01; t_{ac} = 3.44, p < 0.01; t_{bc} = 3.75, p < 0.01$

CONCLUSION

Transvaginal color Doppler is expected to have direct implications for the diagnosis and therapy of ectopic pregnancy, particularly when typical ultrasound morphology (gestational sac and embryonal heart beat) is not present. Since Doppler appears to identify the viability and perhaps the invasiveness of the trophoblast more accurately than any other currently available diagnostic method, it may provide a foundation for a more selective management of ectopic pregnancy.

REFERENCES

1. Sivin, I. and Cooper, T. (1979). IUD use and ectopic pregnancy rates in the United States. *Contraception*, **19**, 151
2. Rubin, G.L., Peterson, H.B., Dorfman, S.F. *et al.* (1983). Ectopic pregnancy in the United States, 1970 through 1978. *J. Am. Med. Assoc.*, **249**, 1725
3. Westrom, L., Bengtsson, L. and Mardh, P.A. (1981). Incidence, trends and risks of ectopic pregnancy in a population of women. *Br. Med. J.*, **282**, 15
4. Dorfman, S.F. (1983). Death from ectopic pregnancy, United States 1979 to 1980. *Obstet. Gynecol.*, **62**, 334
5. Hazenkamp, J.T. (1980). Ectopic pregnancy: diagnostic dilemma and delay. *Int. J. Gynaecol. Obstet.*, **17**, 598
6. Tancer, M.L., Delke, I. and Veridiano, N.P. (1981). A fifteen year experience with ectopic pregnancy. *Surg. Gynecol. Obstet.*, **152**, 179
7. Rottem, S. and Timor-Tritsch, I.E. (1988). Think ectopic. In Timor-Tritsch, I.E. and Rottem, S. (eds.) *Transvaginal Sonography*, p. 125. (New York, Amsterdam, London: Elsevier)
8. Curran, J.W. (1980). Economic consequences of pelvic inflammatory disease in the United States. *Am. J. Obstet. Gynecol.*, **138**, 848
9. Jurković, D., Kurjak, A., Alfirevic, Z. and Klobucar, A. (1988). Ultrasound diagnosis of combined intrauterine and extrauterine pregnancy. *J. Clin. Ultrasound*, **16**, 259
10. Fleischer, A.C., Cartwright, P.S., Di Petro, D.L. and James, A.E. (1985). Sonographic evaluation of ectopic pregnancy. In Sanders, R.C. (ed.) *The Principles and Practice of Ultrasonography in Obstetrics and Gynecology*, p. 399. (Norwalk: Appleton Century Crofts)
11. Kurjak, A., Žalud, I. and Volpe, G. (1990). Conventional B-mode and transvaginal color Doppler in ultrasound assessment of ectopic pregnancy. *Acta Med. Iug.*, **44**, 91
12. Gleicher, N., Giglia, R.V., Deppe, G., Elrad, H. and Friberg, J. (1983). Direct diagnosis of unruptured ectopic pregnancy by real-time ultrasonography. *Obstet. Gynecol.*, **61**, 425
13. Mahony, B.S., Filly, R.A., Nyberg, D.A. and Callen, P.W. (1985). Sonographic evaluation of ectopic pregnancy. *J. Ultrasound Med.*, **4**, 221
14. Kadar, N. and Romero, R. (1982). The timing of a repeat ultrasound examination in the evaluation for ectopic pregnancy. *J. Clin. Ultrasound*, **10**, 211
15. Romero, R., Kadar, N., Jeanty, P., Copel, J.A., Chervenak, F.A., DeCherney, A. and Hobbins, J. (1985). Diagnosis of ectopic pregnancy: value of the discriminatory human chorionic gonadotropin zone. *Obstet. Gynecol.*, **66**, 357
16. Nyberg, D.A., Filly, R.A., Laing, F.C., Mack, L.A. and Zarutskie, P.W. (1987). Ectopic pregnancy: diagnosis by sonography correlated with quantitative hCG levels. *J. Ultrasound Med.*, **6**, 145
17. Van der Lugt, B. and Drogendijk, A. (1985). The disappearance of human chorionic gonadotropin from plasma and urine following induced abortion. *Acta Obstet. Gynecol.*, **64**, 547
18. Rottem, S., Timor-Tritsch, I.E. and Thaler, I. (1990). Assessment of pelvic pathology by high frequency transvaginal sonography. In Chervenak, F.A., Isaacson, G. and Campbell, S. (eds.) *Textbook of Obstetrics and Gynecologic Ultrasound*. (Boston: Little, Brown & Co.) in press
19. Rottem, S., Thaler, I., Levron, J., Peretz, B., Istkowitz, J. and Brandes, J. (1990). Criteria for transvaginal sonographic diagnosis of ectopic pregnancy. *J. Clin. Ultrasound*, **18**, 274
20. Timor-Tritsch, I.E. and Rottem, S. (1987(. Transvaginal ultrasonic study of the fallopian tube. *Obstet. Gynecol.*, **70**, 424
21. DeCrespigny, L.C. (1988). Demonstration of ectopic pregnancy by transvaginal ultrasound. *Br. J. Obstet. Gynaecol.*, **95**, 1253
22. Rampen, A. (1988). Vaginal sonography in ectopic pregnancy: a prospective evaluation. *J. Ultrasound Med.*, **7**, 381
23. Nyberg, D.A., Mack, L.A., Jeffrey, R.B. and Laing, F.C. (1987). Endovaginal sonographic evaluation of ectopic pregnancy: a prospective study. *Am. J. Roentgenol.*, **149**, 1181
24. Shapiro, B., Cullen, M., Taylor, K.J.W. and DeCherney, A.H. (1988). Transvaginal ultrasonography for the diagnosis of ectopic pregnancy. *Fertil. Steril.*, **50**, 425
25. Cacciatore, B., Stenman, U.H. and Ylostalo, P. (1989). Comparison of abdominal and vaginal sonography in suspected ectopic pregnancy. *Obstet. Gynecol.*, **73**, 770
26. Kadar, N. (1989). Transvaginal ultrasound for ectopics. *Fertil. Steril.*, **51**, 909
27. Neiger, R., Bailey, S., Wall, A.M. and Palmisano, G. (1989). Diagnosis of ectopic pregnancy using transvaginal ultrasound scanning. *J. Reprod. Med.*, **34**, 52
28. Dashefsky, S.M., Lyons, E.A., Levi, C.S. and Lindsay, D.J. (1988). Suspected ectopic pregnancy: endovaginal and transvesical ultrasound. *Radiology*, **169**, 181
29. Pannell, R.G., Baltarowich, O.H., Kurtz, A.B., Vilaro, M.M., Rifkin, M.D., Needelman, L., Mitchell, D.G., Mervis, S.A. and Goldberg, B.B. (1987). Complicated first-trimester pregnancies: evaluation with endovaginal ultrasound versus transabdominal technique. *Radiology*, **165**, 79
30. Schurz, B., Eppel, W., Egarter, C., Wnenzel, R. and Reinold, E. (1989). Vaginosonography in gynecology. *Ultraschall Med.*, **10**, 90
31. Funk, A. and Fendel, H. (1988). Improved diagnosis of extrauterine pregnancy by endosonography. *Z. Geburtshilfe Perinatol.*, **192**, 49
32. Taylor, K.J.W., Ramos, I.M., Feyock, A.L., Snower, D.P., Carter, D., Shapiro, B.S., Meyer, W.R. and DeCherney, A.H. (1989). Ectopic pregnancy: Duplex Doppler evaluation. *Radiology*, **173**, 93
33. Kurjak, A., Žalud, I., Alfirevic, Z. and Jurković, D. (1990). The assessment of abnormal pelvic blood flow by transvaginal color Doppler. *Ultrasound Med. Biol.*, in press
34. Kurjak, A., Miljan, M., Jurković, D., Alfirevic, Z. and Žalud, I. (1989). Color Doppler in the assessment of fetomaternal circulation. *Rech. Gynecol.*, **1**, 269
35. Kurjak, A., Jurković, D., Alfirevic, Z. and Žalud, I. (1990). Transvaginal color Doppler imaging. *J. Clin. Ultrasound*, **18**, 227
36. Kurjak, A., Žalud, I. and Crvenković, G. (1989). Transvaginal color Doppler in the assessment of maternal and fetal circulation. Proceedings of International Symposium: *Transvaginal sonography*, Rotterdam, November 2–4, p. 93 (abstr.)
37. Alfirevic, Z. and Kurjak, A. (1990). Transvaginal

color and pulsed wave Doppler in the assessment of blood flow in the first trimester of pregnancy. *J. Perinat. Med.*, **18**, 173
38. Kurjak, A., Schulman, H. and Žalud, I. (1990). Transvaginal color Doppler: new technique for the diagnosis of ectopic pregnancy. *Obstet. Gynecol.*, in press
39. Žalud, I. (1990). Transvaginal color Doppler in the detection of ectopic pregnancy. The *7th Congress of the European Federation of Societies for Ultrasound in Medicine and Biology*, Jerusalem, May 6–11, p. 129 (abstr.)
40. Kurjak, A., Miljan, M. and Žalud, I. (1990). Transabdominal and transvaginal color Doppler in the assessment of fetomaternal circulation during all three trimesters of pregnancy. *J. Reprod. Biol.*, **36**, 240
41. Kurjak, A. and Žalud, I. (1990). Transvaginal color Doppler sonography. In Chervenak, F.A., Isaacson, G. and Campbell, S. (eds.) *Textbook of Obstetrics and Gynecologic Ultrasound.* (Boston: Little, Brown & Co.) in press
42. Taylor, K.J.W. (1989). New techniques for the diagnosis of ectopic pregnancy – transvaginal sonography and Doppler. *ECO Italia'89*, Napoli, October 5–7, p. 135 (abstr.)
43. Taylor K.J.W., Ramos, I., Carter, D., Morse, S.S., Snower, D. and Fortune, K. (1988). Correlation of Doppler ultrasound tumor signals with neovascular morphologic features. *Radiology*, **166**, 57
44. Žalud, I. and Kurjak, A. (1990). The assessment of luteal blood flow in pregnant and non-pregnant women by transvaginal color Doppler. *J. Perinat. Med.*, **18**, 215

10 Tumor Neovascularization

A. Kurjak and I. Žalud

Clearly, there are still many clinical problems in differentiating between benign and malignant adnexal tumors '*in vivo*'. Known features that characterize this difference are the mitotic index and pleomorphism, but they cannot be detected by current imaging methods. However, many malignant tumors have bizarre vascular morphology with abnormal blood flow that can be detected by Doppler ultrasound.

In general, the tumor vasculature consists of (1) vessels recruited from the pre-existing network of the host vasculature, and (2) vessels recruited from the angiogenic response of host vessels to cancer cells[1–3].

Although the tumor vasculature originates from the host vasculature, its organization may be completely different depending upon the tumor type, its growth rate and its location. The architecture is different not only among various tumor types, but also between a spontaneous tumor and its transplants[3]. Macroscopically, the tumor vasculature can be studied in terms of two idealized categories: peripheral and central. In tumors with peripheral vascularization, the centers are usually poorly perfused. In those with central vascularization, one would expect the opposite. In reality, a tumor may consist of many territories, each exhibiting one or the other of these two types of idealized vascular pattern.

Microscopically, the tumor vasculature is highly heterogeneous and does not conform to the standard normal vascular organization (i.e. artery to arteriole to capillaries to postcapillary venule to venule to vein). A key difference between normal and tumor vessels is that the latter are dilated, saccular and tortuous, and may contain tumor cells within the endothelial lining of the vessel wall[3]. In addition, unlike a normal tissue with a relatively fixed route between arterial and venous sides, a tumor may have blood flowing from one venule to another via a series of vessels or directly via an arteriovenous shunt. Furthermore, due to the peculiar nature of the vasculature, the organization of vessels may be different from one location to another and from one time to the next. As a result, one would expect different routes of blood flow in the well-perfused advancing zone, semi-necrotic zone and necrotic zone.

Quantitative studies are limited on the spatial heterogeneities in the tumor perfusion rate as a function of tumor growth (size). Based on perfusion rates, four regions can be recognized in a tumor: (1) an avascular, necrotic region; (2) a semi-necrotic region; (3) a stabilized microcirculation region; and (4) an advancing front. Blood flow rates in necrotic/semi-necrotic regions of tumors are low, while those in non-necrotic regions are variable and substantially higher than surrounding/contralateral host normal tissue[3,4].

In general, as tumors grow larger, their vascular surface area decreases, their intercapillary distance increases and they may develop necrotic foci.

NEOVASCULARIZATION

Great interest and study over the past few years have been focused on the process of neovascularization (angiogenesis). It is the process by which new blood vessels are induced and it occurs in the corpus luteum, in embryogenesis, in tumors and in wound healing. It is, therefore, a process of interest in relation to the applications of Doppler to maternal and fetal medicine.

Solid tumor growth in animals and in man is accompanied by neovascularization. New capillary growth is elicited by a diffusible factor generated by malignant tumor cells. In the absence of neovascularization, most tumors might become dormant at a tiny diameter, perhaps 2–3 mm[5,6].

Folkman and collaborators have demonstrated that the development of an adequate vascular supply is critical to the growth and development of a cancer as well as its metastases[7]. When they injected cancer cells into the non-vascularized anterior chamber of a rabbit eye, nodules greater than 1–2 mm in diameter did not develop. However, the same cells introduced into a vascularized area of the eye developed a stromal blood supply and rapidly grew over the eye[8].

Furthermore, it was possible to show that the neoplastic cells elaborated a soluble tumor angiogenesis factor (TAF) that promoted vascularization of the stroma and permitted the progressive growth of solid tumors[9].

The identification of TAF raised the interesting possibility that cancer growth could be blocked by inhibiting the angiogenesis factor. Indeed, an 'anti-angiogenesis' factor has been extracted from cartilage, a tissue that normally lacks vascularization. Infusion of an extract of cartilage reduces the growth rate of experimental neoplasms in animals[10]. More recently, it was shown that heparin also promotes angiogenesis, and protamine (a heparin antagonist) not only inhibits angiogenesis but also suppresses tumor growth[11].

Thus, it is clear that tumor growth is dependent on blood supply. Often, primary tumors and their metastases outgrow their blood supply and undergo central ischemic necrosis, where the tumor cells are most remote from the vascular supply of the surrounding normal tissue. Thus, with cancers, the primary lesions and their secondary implants may enlarge progressively while the central regions of necroses expand.

There is much evidence that tumors induce new blood vessels. Algire attributed to tumor cells the ability to elicit continuously the growth of new capillary endothelium *in vivo*[12]. Feigin observed vigorous neovascularization in the neighborhood of malignant brain tumors that seemed to be excited or induced by the tumor itself[13]. Many other authors, observing tumor growth in the hamster cheek pouch or in the rabbit ear chamber, have described the intense neovascularization around a growing tumor[14–16]. Chalkley[17] and Warren and Shubik[18] were the first to hypothesize humoral induction of tumor blood vessels.

The original description of arteriographic neovascularity or 'tumor vascularity' is credited to Strickland[19]. He described a tumor vessel as one that was 'deployed seemingly without purpose, keeps to no set course and shows no progressive diminution in caliber'. Gammill *et al.* examined several benign and malignant tumors to evaluate the histological picture shown on angiograms as neovascularity[20]. Newly formed vessels are large capillaries (rather large endothelial-lined capillary-like channels), or sinusoids, and contain no smooth muscle in their walls but, instead, only some fibrous connective tissue. 'Puddling', 'laking' and 'staining' represent the collection of contrast medium in small capillaries or sinusoids. Smooth muscle neither regenerates nor proliferates to any significant degree in adults. If normal smooth muscle can be found in the walls of arteries with a normal architecture, then one can assume that that artery is not new and that it does not represent neovascularity.

Another vascular alteration in malignancy is the presence of arteriovenous shunting represented angiographically by early venous opacification. Arteriovenous shunting may result from a large, focal, direct communication[21], or may reflect multiple microscopic communications in the abnormal tumor microcirculation[22,23]. 'Pooling' or 'laking' results from tumor vessels terminating in amorphous spaces within the tumor parenchyma with or without associated necrosis. A tumor 'stain' or 'blush' reflects the accumulation of contrast material in microscopic spaces. Since the vessels are collapsed in a histological section, it is difficult to know how large these spaces are *in vivo*. Shubik contrasted the appearance of a living tumor – in which the vessels may occupy more than 50% of the tumor volume – with the collapse of the vascular bed in the dying host and the dramatic decrease in tumor vessel size in the fixed specimen[24]. Recent studies with transgenic mice have shown that, for at least one type of cancer, angiogenesis occurs during the transition from hyperplasia to neoplasia[25]. Other workers have recently identified and cloned a new platelet-derived factor that stimulates endothelial cell growth and chemotaxis *in vitro* and angiogenesis *in vivo*[26]. An inhibitor of angiogenesis, which is produced by cells when they are capable of expressing an active cancer suppressing gene, has also been discovered[27,28]. The loss of this inhibitor activity occurs concomitantly with expression of both angiogenesis and tumorigenesis. The process of angiogenesis entails protease activity and cell differentiation, proliferation and migration. The vascular morphology of one tumor differs from another and is determined to some extent by the growth pattern of cancer cells. Quantitative morphometric studies in induced animal tumors show that vascular volume, length and surface area increase during the early stages of tumor growth, and then decrease after the onset of necrosis. The frequency of large-diameter vessels increases in the later stages of growth[26]. New blood vessels and vascular channels in a tumor arise from older, pre-existing vessels. Tumor vessels have a relative paucity of smooth muscle in their walls in comparison to their caliber. Since most of the resistance to flow resides at the level of the muscular arterioles, vessels deficient in these muscular elements present diminished resistance to flow and thereby receive a larger volume flow than vessels with a high impedance.

Obviously, microcirculation plays an important role in the growth, metastasis, detection and treatment of tumors. Transabdominal pulsed Doppler imaging offers a view of the surrounding anatomy and evidence of blood flow in major pelvic vessels, but not in the microcirculation[29,30]. By contrast, transvaginal color Doppler offers a qualitative picture of blood flow in the vascular system in relation to the surrounding anatomy (Figures 10.1 and 10.2)[31–37].

TISSUE CHARACTERIZATION

The term 'tissue characterization' covers a range of meanings from qualitative assessment to scientific measurement[38]. Many attempts have been made in the past 20 years to derive quantitative information relating to tissue structure from the returned ultrasound beam. Various acoustical properties of tissues – absorption, speed, dispersion density, compressibility, bulk modulus, scattering characteristics, attenuation – might be probed by 'ultrasonic telehistology'[39].

Other conceptual developments are the non-invasive detection of blood flow by Doppler[40], tissue motion[41] and image texture[42] as possible indicators of histology and pathology.

There are at least seven characteristics that can be measured by ultrasound:

(1) the attenuation of ultrasound[43, 44],
(2) the speed of ultrasound[45],
(3) the acoustic impedance[46],
(4) the scattering characteristics[47],
(5) the non-linearity of signal propagation[48],
(6) the motion of tissue[49],
(7) the detection and quantification of blood flow by techniques based on the Doppler effect[40, 50].

Despite all these efforts, only image inspection and Doppler techniques have shown progress toward the concept of telehistology and it is really only through them that tissue characterization has achieved routine clinical application.

Tissue characterization by conventional Doppler equipment

Wells *et al.*[40] and Burns *et al.*[50] documented the presence of abnormal flow spectra around the periphery of malignant tumors. These results were confirmed by several groups[51–58]. The abnormal flow signals consisted of an increase in signal amplitude when compared with the contralateral normal side (corresponding to a greater number of moving cells within the beam); an increase in peak systolic flow; a characteristic distribution of the Doppler spectrum, showing a predominance of high-power, low-frequency elements; and high diastolic shift (in some tumors the systolic/diastolic variation is absent).

The major difficulty with both continuous wave Doppler (CWD) and pulsed wave Doppler (PWD) is the need for operator expertise and time to search the periphery of the tumor and to detect one of the limited number of localized abnormal vessels or shunts.

Tissue characterization by color Doppler

Color Doppler imaging is the latest addition to the many ultrasound techniques available to the clinician. It is best regarded as an extension of pulsed wave Doppler and conventional two-dimensional imaging. Color Doppler imaging is a type of pulsed wave Doppler. Both are specific in range information, i.e. the position of the Doppler shift information is known. However, color Doppler imaging can be explained by comparing its principles with pulsed wave Doppler. To begin with, in conventional pulsed wave Doppler, a sample volume is positioned from a small discrete area. In color Doppler imaging, Doppler shifted information is retrieved from multiple sample volumes along many two-dimensional scan lines. So, color Doppler imaging is an area analysis which may be all or part of the two-dimensional image. Each sample volume (gate) in color Doppler imaging determines a mean velocity at that point, and all gates are displayed. So, the overall view of the flow situation within the two-dimensional image is obtained. Technologically, the measurement is, in some scans, performed using measurements other than Doppler-frequency shift, but the results are equivalent to Doppler measurements.

Both the pulsed wave Doppler and color Doppler imaging display contain information about *direction* of flow and *velocity* of flow. In pulsed wave Doppler, the *direction* is represented by displaying a spectral waveform *above* a zero line for flow *away from* the transducer. The *velocity* in pulsed wave Doppler is represented by the *amplitude* of the waveform relative to the zero line. In color Doppler imaging, both directions and velocities are represented on the color image. *Directions* are usually indicated by two color scales, reds —> yellows for flow *towards*, and blues —> greens for flow *away* (Figure 10.3).

Both pulsed wave Doppler and color Doppler imaging are based upon the laws of the Doppler equation. Hence, color Doppler imaging has all the limitations of the pulsed wave Doppler, e.g. dependence on the angle of flow applies to both systems. The best Doppler signals are obtained when interrogation is at low angle to flow, while the best two-dimensional imaging angles are high.

To summarize, it may be useful to compare the fundamental properties of the different forms of ultrasound imaging:

- Two-dimensional imaging gives anatomical information.

- Pulsed wave Doppler spectral imaging gives functional information.

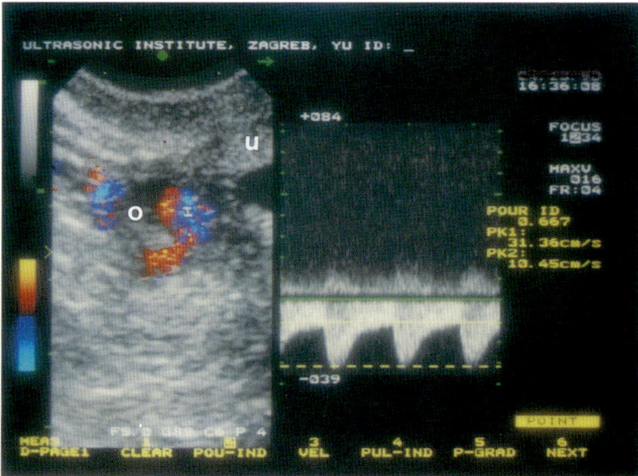

Figure 10.1 Transvaginal color Doppler shows abundant blood flow at the ovarian periphery. Pulsed Doppler (right) shows moderate velocity and moderate resistance of blood flow. These are typical normal findings of the terminal branch of the uterine artery. O = ovary ; u = uterus

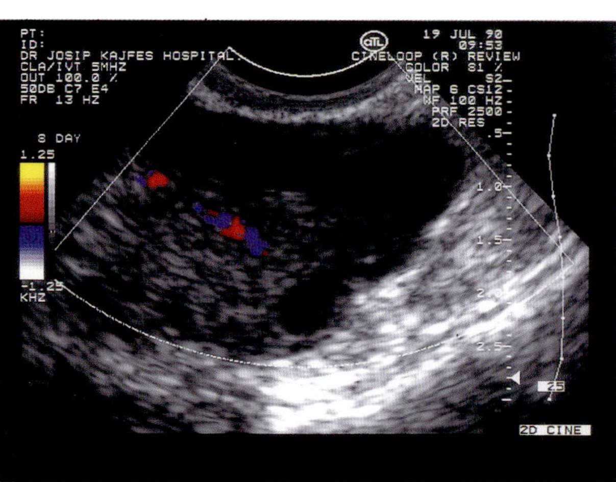

Figure 10.2 A slightly enlarged ovary detected by B-mode ultrasound. Superimposed color Doppler shows highly vascularized ovarian tissue in the luteal part of the menstrual cycle. Even small newly formed vessels can be studied

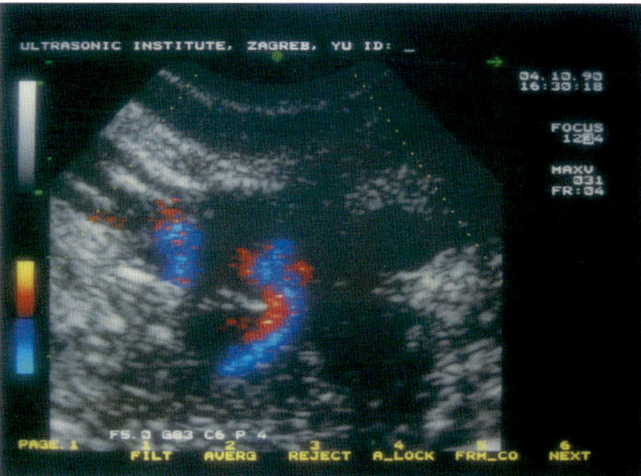

Figure 10.3 Transvaginal color Doppler of the uterine vessels. The red color indicates flow towards the probe and blue color indicates flow away from the probe. The brightness of color corresponds to the velocity of blood flow

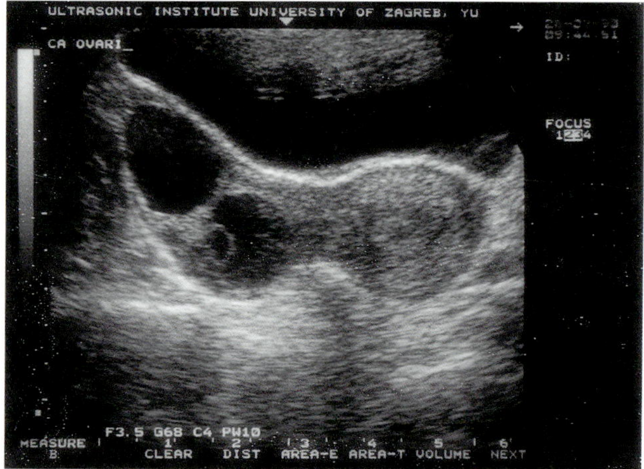

Figure 10.4 A transabdominal scan of the right adnexal region. A large predominantly multicystic adnexal mass can be seen. Morphological features suggest ovarian cancer

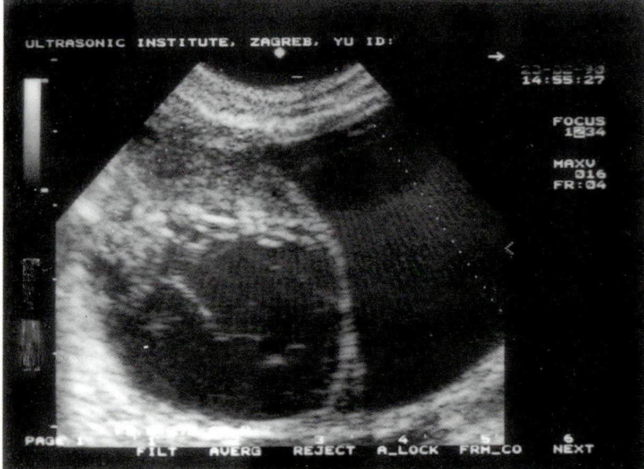

Figure 10.5 The transvaginal scan of the same patient also supports the suspicion of ovarian malignancy

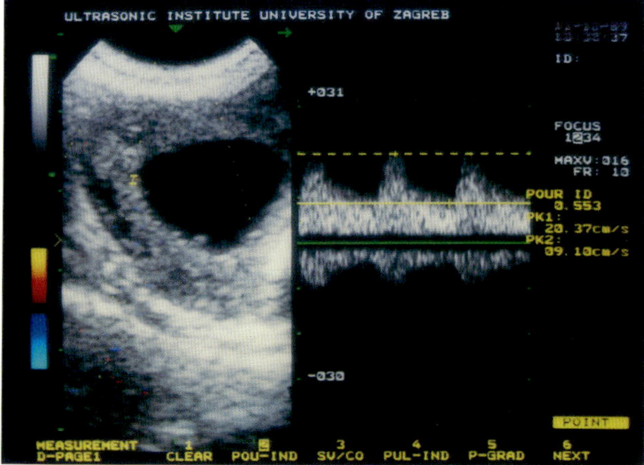

Figure 10.6 The same patient. Transvaginal color Doppler shows blood flow in the septal part of the adnexal mass. The waveform analysis (right) indicates a benign finding (RI = 0.553). The benign nature of the tumor was finally confirmed by histopathology

- Duplex imaging gives anatomical and functional information.
- Color Doppler imaging gives functional information at many anatomical positions simultaneously.

Thus, the main advantage of color Doppler imaging is the ability to screen large areas rapidly to obtain *anatomical* and *functional* information.

In assessing the unique abilities of color Doppler, the overall advantages and clinical utility, in addition to ease of application, must be considered. Firstly, the format of the information must be understood by reference to the basic physics and methodology. There is no doubt that an understanding of Doppler principles is enhanced by the use of color Doppler imaging. The learning time for new users is considerably reduced as a result. Also, the increased information due to large *area* interrogation leads to shorter examination time. This aspect alone is sufficient reason to claim that the color Doppler imaging technique can be applied easily and routinely for the examination of most fetal vessels which are currently the subject of interest, and also the vessels of the umbilical cord. Since the main vessels, as previously mentioned, can be visualized in color almost instantaneously, the optimum Doppler signals (pulsed wave spectral signals) can be obtained by choosing an appropriate angle.

By these techniques, the spectral waveform 'signature' of each vessel can be rapidly appraised and changes in these can be recognized in conditions of pathology and/or changes in physiology.

In a similar manner, the 'signature' waveforms of maternal circulation (e.g. uterine arteries, spiral arteries) may be established and monitored. Color Doppler imaging has enabled easier recognition of these vessels and hence has increased the reliability related to repeatability.

In summary, the popularly called color Doppler is in fact a two-dimensional pulsed Doppler system with color coding for blood flow velocities. It has all the basic limitations and advantages of the pulsed Doppler systems, i.e. limitations for measurement of high blood velocities deep in the body and a well-defined axial resolution, respectively. The main advantage is that it measures the two-dimensional distribution of flow in an area making the orientation much easier and, in fact, making possible the search for disturbed flow in small vessels. This feature is of utmost importance in the examination of tumors where the distribution of blood vessels is never known in advance. It is this search ability which makes it potentially so powerful in solving some of the diagnostic problems so far unsolved.

The problems arising in the measurement of high blood flow velocity in the heart are much less when measuring the flow in the female genital organs. On the other hand, orientation and finding the blood vessels in this area are difficult and, with conventional B-mode systems, often impossible. This renders the two-dimensional Doppler method the only practical one for assessment of the flow in the female genital tract. The major difficulty with both continuous wave and pulsed Doppler is the need for operator expertise and time to search the periphery of the tumor and detect one of the limited number of localized shunts. Color Doppler, by showing a two-dimensional display of vascularity, should expedite the examination by displaying simultaneously the presence of all the vessels, which can then be interrogated for quantitative analysis by pulsed Doppler.

Taylor *et al.*[55] and Hata *et al.*[59, 60] were the first to apply pulsed Doppler and transabdominal color Doppler in the study of female pelvic circulation. In all cases of endometrial carcinoma, ovarian carcinoma and trophoblastic disease, typically abnormal flows were observed by the Japanese team. However, all cases of cervical carcinoma with abnormal flows were stage II-B and above. They concluded that Doppler ultrasound is a pertinent diagnostic tool that can be used to observe changes in tumor vascularity in gynecological malignancies before and after treatment.

TRANSVAGINAL COLOR DOPPLER

Kurjak *et al.*[31–37, 61–68] and Žalud and Kurjak[69] were the first to report that transvaginal color flow imaging can be used in the assessment of pelvic circulation, and to differentiate between benign and malignant pelvic tumors (Figures 10.4–10.14).

Using transvaginal color Doppler, Bourne *et al.* showed that the absence of intratumoral neovascularization and a normal (high) pulsatility index can be used to exclude the presence of invasive primary ovarian cancer[70]. The early recognition of ovarian cancer is the only approach to achieve a reduction in the mortality. Transvaginal color flow mapping may be used to identify potentially malignant ovarian masses and help to elucidate the early stages of tumorigenesis. The routine application of this new technique will enable us to develop a screening program based on ultrasonography.

In their preliminary report, Hata *et al.* also confirm that transvaginal color Doppler is expected to be an important diagnostic method for assessing blood flow in physiological conditions of the pelvis[71]. However, a transesophageal probe was used in their study.

Intratumoral blood flow, displayed on transvaginal color Doppler images, indicated that there is a flow rapid enough to be detected. The presence of

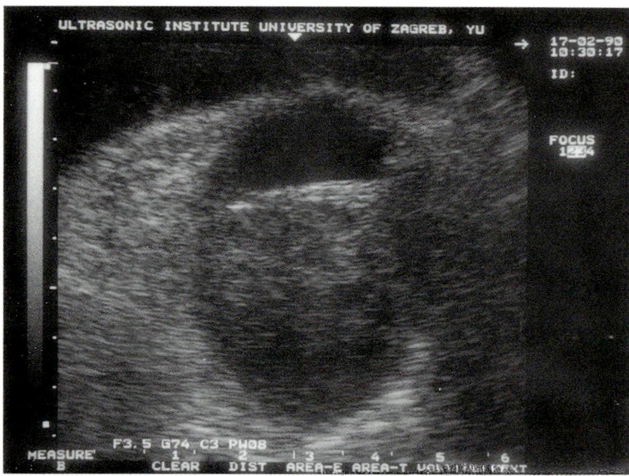

Figure 10.7 Transabdominal ultrasound of the solid adnexal mass. The sonographic appearance is not highly suspicious of malignancy

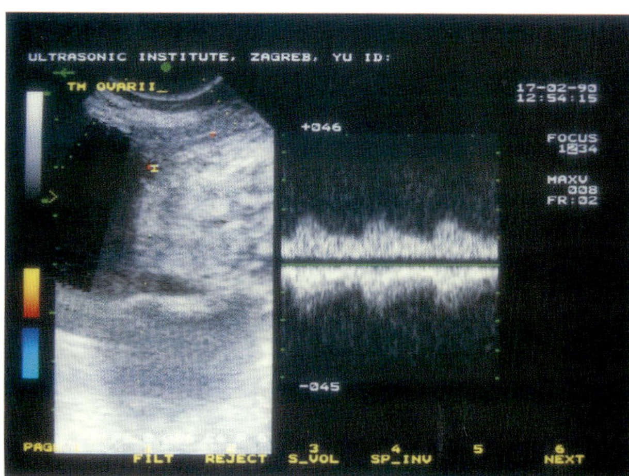

Figure 10.8 The same patient examined by transvaginal color Doppler. Small vessels are detected on the border of the cystic and solid parts of the tumor. Pulsed Doppler shows a small systolic–diastolic variation of blood flow. A final diagnosis of ovarian cystadenocarcinoma was made

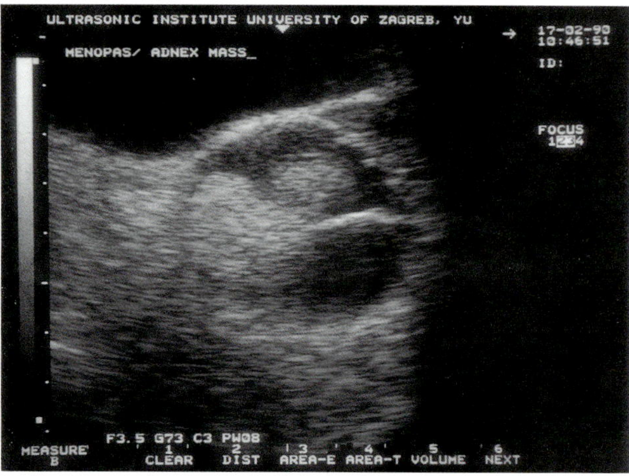

Figure 10.9 A mixed adnexal mass detected in a postmenopausal woman by transabdominal ultrasound. The morphological study supports the suspicion of ovarian malignancy

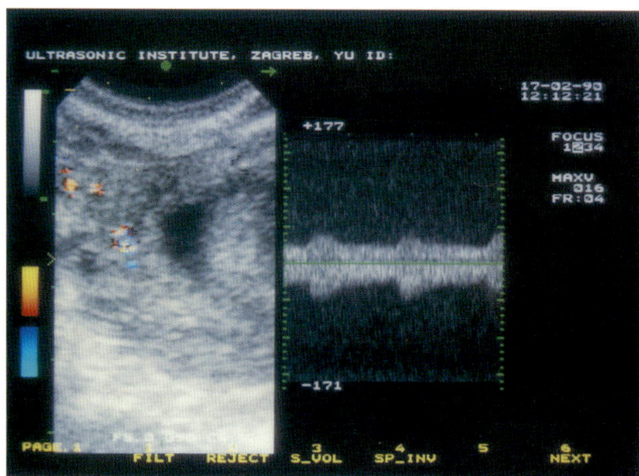

Figure 10.10 The same patient. Small tumor vessels are detected by transvaginal color Doppler. Pulsed Doppler (right) analysis shows low velocity and very low resistance of blood flow. The Doppler finding also strongly supported the diagnosis of ovarian malignancy. The final diagnosis was ovarian cystadenocarcinoma

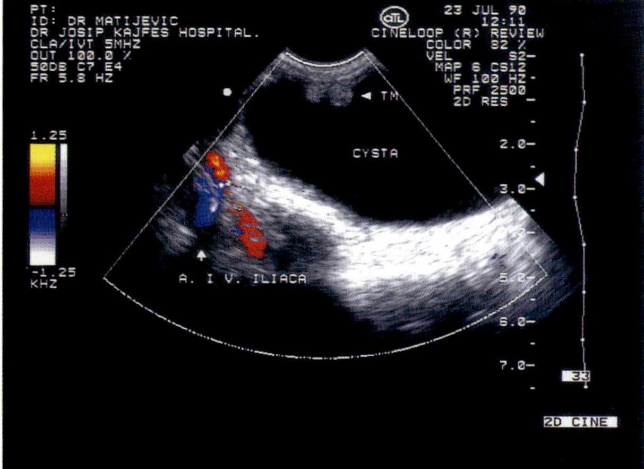

Figure 10.11 A cystic adnexal tumor with papillary proliferation. Neovascularization could not be documented. The benign nature of the tumor was proved by histopathology

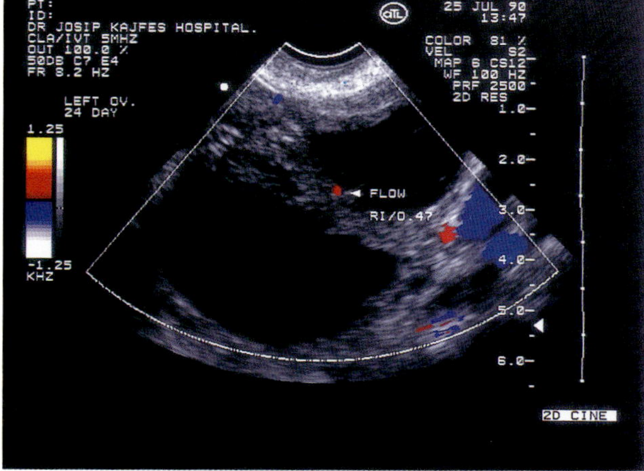

Figure 10.12 A multilocular cystic adnexal mass. Blood flow was visualized in the septa and the RI was 0.47

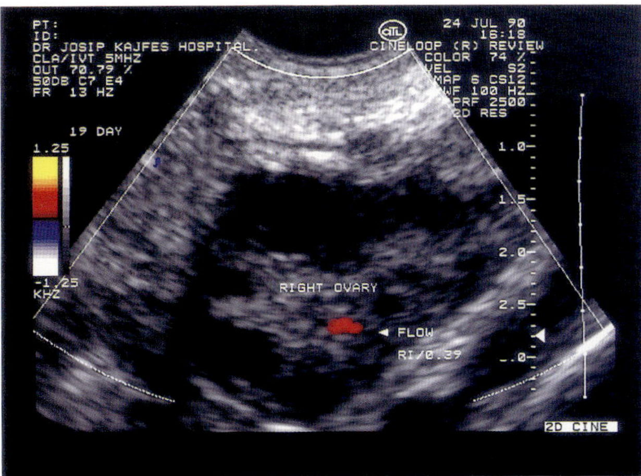

Figure 10.13 Another example of ovarian neovascularization

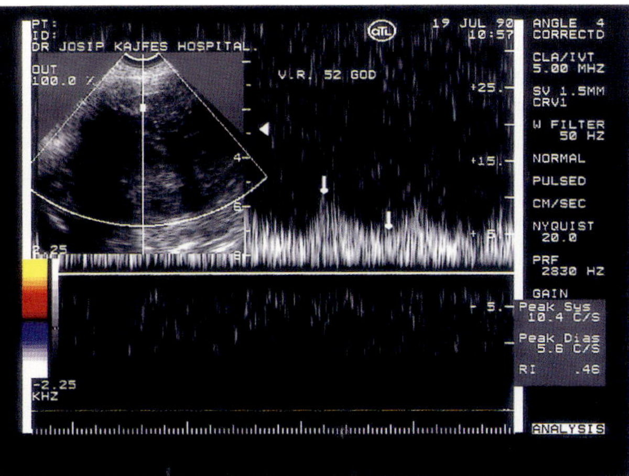

Figure 10.14 Uterine myoma. Blood flow was detected on the tumor periphery. Pulsed Doppler analysis (right) shows increased diastolic flow (RI = 0.46)

arteriovenous communications should be an important factor that produces sufficient velocity above the minimal threshold on color Doppler imaging. When tumoral blood flow is not visualized on transvaginal color Doppler examination, the following factors should be taken into consideration:

(1) There is no blood flow due to the lack of newly formed vessels characteristic of malignant tumor.

(2) The velocity of flow may be too slow to exceed the minimal threshold for measurement with current color Doppler systems.

(3) Intratumoral blood flow is non-uniform and turbulent, and detectable blood flow on color Doppler imaging is distributed in certain regions of a tumor. It sometimes requires considerable effort to obtain a good angle of incidence to the target for color flow imaging.

REFERENCES

1. Gulino, P.M. (1975). Extracellular compartments of solid tumors. In Becker, H. (ed.) *Cancer*, p. 327. (New York: Plenum Press)
2. Folkman, J. (1985). Tumor angiogenesis. *Adv. Cancer. Res.*, **43**, 175
3. Jain, R.K. (1988). Determination of tumor blood flow: a review. *Cancer Res.*, **48**, 2641
4. Jain, R.K. and Ward-Hartley, K. (1984). Tumor blood flow – characterization, modifications and role in hyperthermia. *Trans. Sonics Ultrasonics*, **31**, 504
5. Jain, R.K. (1987). Transport of molecules across tumor vasculature. *Cancer Metastasis Rev.*, **6**, 559
6. Jain, R.K. and Baxter, L.T. (1988). Mechanisms of heterogeneous distribution of monoclonal antibodies and other macromolecules in tumors: significance of elevated interstitial pressure. *Cancer Res.*, **48**, 7022
7. Folkman, J. (1963). Growth and metastasis of tumor in organ culture. *Cancer*, **16**, 453
8. Gimbrone, M.A. (1972). Tumor dormancy *in vivo* by prevention of neovascularization. *J. Exp. Med.*, **136**, 261
9. Folkman, J. and Cotran, R.S. (1976). Relation of vascular proliferation to tumor growth. *Int. Rev. Exp. Pathol.*, **16**, 207
10. Langer, R. (1980). Control of tumor growth in animals by infusion of an angiogenesis inhibitor. *Proc. Natl. Acad. Sci. USA*, **77**, 4331
11. Taylor, S. and Folkman, J. (1982). Protamine is an inhibitor of angiogenesis. *Nature (London)*, **297**, 307
12. Algire, G.H. and Chalkley, H.W. (1945). Vascular reactions of normal and malignant tissues *in vivo*. I. Vascular reactions of mice to wounds and to normal and neoplastic implants. *J. Natl. Cancer Inst.*, **6**, 73
13. Feigin, I., Allen, L.B. and Lipkin, O. (1958). The endothelial hyperplasia of the cerebral blood vessels with brain tumors and its sarcomatous transformation. *Cancer*, **11**, 264
14. Goodall, C.M., Feldman, R., Sanders, A.G. and Shubik, P. (1965). Vascular patterns of four transplantable tumors in the hamster (*Mesocricetus auratus*). *Angiology*, **16**, 622
15. Ide, A.G., Baker, N.H. and Warren, S.L. (1939). Vascularization of the brown-pearce rabbit epithelioma transplant as seen in the transparent ear chamber. *Am. J. Roentgenol.*, **42**, 891
16. Warren, B.A. and Shubik, P. (1966). The ultrastructure of capillary sprouts induced by melanoma transplants in the golden hamster. *J.R. Micro. Soc.*, **86**, 177
17. Chalkley, H.W. (1948). Comments on Algire, G.H. and Lagallais, F.Y. Growth and vascularization of transplanted mouse melanoma. Special publications, *N.Y. Acad. Sci.*, **4**, 164
18. Warren, B.A. and Shubik, P. (1966). The growth of blood supply to melanoma transplants in the hamster cheek pouch. *Lab. Invest.*, **15**, 464
19. Strickland, B. (1959). The value of arteriography in the diagnosis of bone tumors. *Br. J. Radiol.*, **32**, 705
20. Gammill, S.L., Shipkey, F.H., Himmelfarb, E.H., Parvey, L.S. and Rabinowitz, J.G. (1976). Roentgenology–pathology correlation study of neovascularization. *Am. J. Radiol.*, **126**, 376
21. Okuda, K., Musha, H. and Yamasaki, T. (1977). Angiographic demonstration of intrahepatic arterioportal anastomoses in hepatocellular carcinoma. *Radiology*, **122**, 53
22. Tegtmeyer, C.J. (1982). Angiography of bones, joints and soft tissues. In Abrams, H.L. (ed.) *Angiography*, p. 1939. (Boston: Little, Brown &Co.)
23. Kadir, S. (1986). Neoplasms and neoplasm-like conditions of the extremities, shoulders and pelvis. In

Bom, N. (ed.) *Diagnostic Angiography,* p. 323. (Philadelphia: Saunders)

24. Shubik, P. (1982). Vascularization of tumors: a review. *J. Cancer Res. Clin. Oncol.*, **103**, 211
25. Folkman, J., Watson, K., Ingber, D. and Hanahan, D. (1989). Induction of angiogenesis during the transition from hyperplasia to neoplasia. *Nature (London)*, **339**, 58
26. Jain, R.K. and Ward-Hartley, K.A. (1987). Dynamics of cancer cell interaction with microvasculature and interstitium. *Biorheology*, **24**, 117
27. Folkman, J. (1972). Anti-angiogenesis: new concept for therapy of solid tumors. *Ann. Surg.*, **175**, 409
28. Folkman, J., Merler, E., Abernathy, C. and Williams, G. (1971). Isolation of a tumor factor responsible for angiogenesis. *J. Exp. Med.*, **133**, 275
29. Taylor, K.J.W., Burns, P.N., Wells, P.N.I. and Conway, D.I. (1985). Ultrasound Doppler flow studies of the ovarian and uterine arteries. *Br. J. Obstet. Gynaecol.*, **92**, 240
30. Long, M.G., Boulbee, J.E., Hanson, M.E. and Begent, R.H.J. (1989). Doppler time velocity waveform studies of the uterine artery and uterus. *Br. J. Obstet. Gynaecol.*, **96**, 588
31. Kurjak, A., Žalud, I., Jurković, D., Alfirevic, Z. and Miljan, M. (1989). Transvaginal color Doppler for the assessment of pelvic circulation. *Acta Obstet. Gynecol. Scand.*, **68**, 131
32. Kurjak, A., Žalud, I., Alfirevic, Z. and Jurković, D. (1990). The assessment of abnormal pelvic blood flow by transvaginal color Doppler. *Ultrasound Med. Biol.*, in press
33. Kurjak, A., Jurković, D., Alfirevic, Z. and Žalud, I. (1990). Transvaginal color Doppler imaging. *J. Clin. Ultrasound*, **18**, 227
34. Kurjak, A. (1989). Transvaginal color Doppler in the assessment of pelvic circulation. *Jpn. J. Med. Ultrasound*, **16** (Suppl.II), 1
35. Kurjak, A. and Jurković, D. (1989). Transvaginal color Doppler in the assessment of pelvic masses. Proceedings of *6th World Congress of In Vitro Fertilization and Alternate Assisted Reproduction*, Jerusalem, Abstr. p.26
36. Kurjak, A. and Žalud, I. (1990). Transvaginal color Doppler in the characterization of pelvic masses. *Ultraschall Med.*, in press
37. Kurjak, A. and Žalud, I. (1990). Transvaginal color Doppler. In Chervenak, F.A., Isaacson, G. and Campbell, S. (eds.) *Textbook of Ultrasound in Obstetrics and Gynecology*. (Boston: Little, Brown and Company) in press
38. Chivers, R.C. (1981). Tissue characterization. *Ultrasound Med. Biol.*, 7, 1
39. Chivers, R.C. and Hill, C.R. (1973). An approach to telehistology: ultrasonic scattering by tissues. *Br. J. Radiol.*, **46**, 567
40. Wells, P.N.T., Halliwell, M., Skidmore, R., Webb, A.J. and Woodcock, J.P. (1977). Tumor detection by ultrasonic Doppler blood flow signals. *Ultrasonics*, **15**, 231
41. Tristam, M., Barbosa, D.C., Cosgrove, D.O., Nassiri, D.K., Bamber, J.C. and Hill, C.R. (1986). Ultrasonic study of *in vivo* kinetic characterization of human tissues. *Ultrasound Med. Biol.*, **12**, 927
42. Lerski, R.A., Smith, M.J., Morley, P., Barnett, E., Mills, P.R., Watkinson, G. and MacSween, R.N.M. (1982). Discriminant analysis of ultrasonic texture data in diffuse alcoholic liver disease. *Ultrason. Imaging*, **3**, 164
43. Mountford, R.A. and Wells, P.N.T. (1972). Ultrasonic liver scanning: the qualitative analysis of the normal A-scan. *Phys. Med. Biol.*, **17**, 14
44. Wells, P.N.T. (1980). Propagation of ultrasonic waves through tissue. In Fullerton, G.D. and Zagzebski, J.A. (eds.) *Medical Physics of CT and Ultrasound*, p. 367. (New York: American Institute of Physics)
45. Greenleaf, J.F., Kenue, S.K., Rajagopolan, B., Bahn, R.C. and Johnson, S.A. (1980). Breast imaging by ultrasonic computer-assisted tomography. In Matherell, A.F. (ed.) *Acoustical Imaging*, Vol.8, p. 599. (New York: Plenum Press)
46. Herment, A., Peronneau, P. and Vayse, M. (1979). A new method of obtaining an acoustic impedance profile for characterization of tissue structures. *Ultrasound Med. Biol.*, **5**, 321
47. Dickinson, R.J. (1986). Reflection and scattering. In Hill, C.R. (ed.) *Physical Principles of Medical Ultrasound*, p. 225. (Chichester: Ellis Harwood)
48. Law, W.K., Frizzell, L.A. and Dunn, F. (1985). Determination of the nonlinearity parameter B/A of biological media. *Ultrasound Med. Biol.*, **11**, 307
49. Tristam, M., Barbosa, D.C., Cosgrove, D.O., Nassiri, D.K., Bamber, J.C. and Hill, C.R. (1986). Ultrasonic study of *in vivo* kinetic characterization of human tissues. *Ultrasound Med. Biol.*, **12**, 927
50. Burns, P.N., Halliwell, M., Wells, P.N.T. and Webb, A.J. (1982). Ultrasonic Doppler studies of the breast. *Ultrasound Med. Biol.*, **8**, 127
51. Mountford, R.A. and Atkinson, P. (1979). Doppler ultrasound examination of pathologically enlarged lymph nodes. *Br. J. Radiol.*, **52**, 464
52. Minasian, M. and Bamber, J.C. (1982). A preliminary assessment of an ultrasonic Doppler method for the study of blood flow in human breast cancer. *Ultrasound Med. Biol.*, **8**, 357
53. Jellins, J., Kossoff, G., Boyd, J. and Reeve, T.S. (1983). The complementary role of Doppler to the B-mode examination of the breast. *J. Ultrasound Med.*, **2**, 10
54. Srivastava, A., Huges, L.E., Woodcock, J.P. and Laider, P. (1989). Vascularity in cutaneous melanoma detected by Doppler sonography and histology: correlation with tumor behavior. *Br. J. Cancer*, **59**, 89
55. Taylor, K.J.W., Ramos, I., Morse, S.S., Fortune, K., Hammers, L. and Taylor, C.R. (1987). Focal liver masses: differential diagnosis with pulsed Doppler ultrasound. *Radiology*, **164**, 643
56. Taylor, K.J.W. and Morse, S.S. (1988). Doppler defects vascularity of some malignant tumors. *Diagn. Imaging*, **10**, 132
57. Taylor, K.J.W., Ramos, I., Carter, D., Morse, S.S., Snower, D. and Fortune, K. (1988). Correlation of Doppler ultrasound tumor signals with neovascular morphologic features. *Radiology*, **166**, 57
58. Kujipers, D. and Jaspers, R. (1989). Renal masses: differential diagnosis with pulsed Doppler ultrasound. *Radiology*, **170**, 59
59. Hata, T., Hata, K., Yamane, Y. and Kitao, M. (1988). Real-time two-dimensional and pulsed Doppler ultrasound detection of intrapelvic neoplastic tumor and abnormal pathogenic changes: preliminary report. *J. Cardiovasc. Ultrasonog.*, **7**, 135
60. Hata, H., Hata, K., Senoh, D., Makihara, K., Aoki, S., Takamiya, O. and Kiato, M. (1989). Doppler ultrasound assessment of tumor vascularity in gynecological disorders. *J. Ultrasound Med.*, **8**, 309

61. Kurjak, A. and Žalud, I. (1989). Early diagnosis of ovarian tumors: transvaginal color Doppler ultrasound. *ECO Italia '89*, Napoli, October 2–7, Abstr. p. 189
62. Kurjak, A. and Žalud, I. (1990). Transvaginal colour flow imaging and ovarian cancer. *Br. Med. J.*, **300**, 330
63. Kurjak, A., Žalud, I. and Crvenković, G. (1989). Transvaginal color Doppler in the assessment of maternal and fetal circulation. Proceedings of International Symposium: *Transvaginal Sonography*, Rotterdam, November 2–4, Abstr. p. 93
64. Kurjak, A., Žalud, I. and Funduk-Kurjak, B. (1990). Transvaginal color Doppler in the assessment of uterine perfusion in infertile women. The *7th Congress of the European Federation of Societies for Ultrasound in Medicine and Biology*, Jerusalem, May 6–11, Abstr. p. 5
65. Kurjak, A. (1990). Transvaginal color Doppler in the detection of ovarian malignancy. The *7th Congress of the European Federation of Societies for Ultrasound in Medicine and Biology*, Jerusalem, May 6–11, Abstr. p. 91
66. Kurjak, A., Žalud, I. and Grljusic, V. (1990). Transvaginal color Doppler in the blood flow studies of postmenopausal women. The *7th Congress of the European Federation of Societies for Ultrasound in Medicine and Biology*, Jerusalem, May 6–11, Abstr. p. 91
67. Kurjak, A. and Žalud, I. (1990). Transvaginal color Doppler. In Kurjak, A. (ed.) *Handbook of Ultrasound in Obstetrics and Gynecology*, Vol. 2, pp. 305–12. (Boca Raton, Florida: CRC Press)
68. Kurjak, A., Žalud, I. and Crvenković, G. (1990). The assessment of pelvic circulation by transvaginal color Doppler. *Jpn. J. Med. Ultrasound*, **17**, 116
69. Žalud, I. and Kurjak, A. (1990). The assessment of luteal blood flow in pregnant and non-pregnant women by transvaginal color Doppler. *J. Perinat. Med.*, **18**, 215
70. Bourne, T., Campbell, S., Steer, C., Whitehead, M.I. and Collins, W.P. (1989). Transvaginal colour flow imaging: a possible new screening technique for ovarian cancer. *Br. Med. J.*, **299**, 1367
71. Hata, T., Hata, K., Senoh, D., Makihara, K., Aoki, S., Takakiya, O. and Kitao, M. (1989). Transvaginal Doppler color flow mapping. *Gynecol. Obstet. Invest.*, **27**, 217

11 Adnexal Masses

A. Kurjak, I. Žalud and H. Schulman

The ovary is complex in its embryology, histology, steroidogenesis and potential for developing malignancy. Furthermore, it is made up of germinal epithelium, germ cells of the gonadal stroma, and mesenchymal cells – each with its own potential to form a tumor. In addition, the ovary is unique in that it not only gives rise to a great variety of malignancies but it is also a favorite site for metastasis from other organs.

Despite progress in chemotherapy and surgery, ovarian carcinoma remains the most lethal of the gynecological malignancies and its incidence continues to rise with age[1]. The American Cancer Society estimates that roughly 20 000 new cases of the disease will be diagnosed in the US in 1989 and that about 60% of those affected will die of the disease. In fact, ovarian cancer is now expected to affect 1 in 70 women in their lifetime[2]. Despite the histological complexity of the ovary, the most frequent cancers are of epithelial origin, accounting for nearly 90% of all cases reported[3]. While ovarian cancers generally begin in the ovaries, they frequently spread throughout the entire peritoneal cavity by direct peritoneal dissemination. Thus, the disease is considered to be a loco-regional disease, with distant dissemination beyond the abdomen generally a late event. The FIGO stage of the disease is a critical determinant of the probability of survival. With stage Ia disease (limited to the ovary after careful staging by a full abdominal exploration), survival in well-differentiated tumors is nearly 90%. When the disease is stage III (primary tumor plus: intraperitoneal metastases outside the pelvis and/or positive retroperitoneal nodes – or pelvic primary disease with bowel involvement) only a 10–20% 5-year survival is seen[2,4].

Unfortunately, a preinvasive stage of ovarian cancer is still unknown. Therefore, the prognosis for ovarian cancer has remained unchanged for 30 years. Attempts to detect early localized ovarian cancer have not yet been successful and the early diagnosis of ovarian cancer is still a matter of chance rather than a triumph of any scientific method. Means of early diagnosis are extremely limited. It is the capacity of cancer to grow and disseminate that makes early diagnosis so important. A cancerous tumor 1 cm^3 in size contains about a billion cells, each potentially capable of originating a new focus of cancer. Discovery of cancer when the tumor is considerably smaller than this should materially increase cure rates[5].

SCREENING TESTS FOR OVARIAN CANCER

Although the outcome may be improved by the introduction of new treatment regimens, the best possible approach for further reduction in mortality is the development of efficacious screening for malignant ovarian tumors. Screening has been defined as the 'identification, among apparently healthy individuals, of those who are sufficiently at risk of a specific disorder to justify a subsequent diagnostic test or procedure, or in certain circumstances, direct preventive action'[6]. This definition highlights two important differences between screening tests and diagnostic tests. Firstly, in screening there is no intention to make a diagnosis or to offer therapeutic intervention solely on the basis of a positive result. Secondly, screening tests are applied to apparently healthy individuals rather than to those seeking attention for a medical problem. In screening, there is an obligation not to initiate any action unless the full consequences of doing so are known. This requires rigorous evaluation[7].

In an effort to diagnose ovarian malignancy, there has been important progress during the last decade. Transvaginal sonography and immunological detection of CA 125 antigen provide a detailed depiction of the ovaries, raising the possibility that screening for early-stage ovarian cancer may now be feasible. Consequently, there have been a number of studies about the performance characteristics of potential screening tests[8]. It is hoped that screening and early detection of ovarian cancer with these two diagnostic modalities will prove as efficacious as the Pap smear has been in screening for cervical cancer. Cervical cancer is effectively screened with a Pap smear, which detects not only non-invasive cancer but also precursors to the neoplasm. Since the incidence of carcinoma of the ovary is higher than cancer of the cervix, screening for ovarian cancer could be practical and indicated medically. However, the ideal screening test for ovarian cancer should detect disease in a premalignant phase and, hence, provide a method for the prevention of invasion. Unfortunately, there is still no well-defined precancerous lesion of the ovary analogous to cervical intraepithelial neoplasia or atypical endometrial hyperplasia. On the other hand, most of us know quite well that a benign ovarian cyst may sometimes be the site of malignant transformation when a small malignant focus develops in an otherwise benign neoplasm.

Indeed, any consideration for screening of early-stage ovarian cancer must take into account that the natural history of ovarian malignancy, and in particular, the length of the preclinical phase, is unclear. If the preclinical phase is short, the screening interval will have to be too frequent to be of any value in clinical practice.

There are a number of restrictions on the nature of a screening test for ovarian cancer. Screening must be performed by techniques which either demonstrate a change in the structure of the ovary (e.g. size, morphology) or an alteration in ovarian function (e.g. as reflected by the release of metabolites into the peripheral circulation). Furthermore, the changes detected must be highly specific for the presence of malignancy, because the incidence of ovarian cancer is relatively low and the consequence of a positive screening test is surgical operation. A test with a positive predictive value of less than 10% would be unacceptable in the context of screening for ovarian cancer where the diagnostic procedure is laparoscopy or laparotomy[9]. As the most appropriate selection criteria for screening remains age, a level of specificity of the order of 99.6% is an essential requirement of any screening test for ovarian cancer. Unfortunately, the level of sensitivity required of a screening test for ovarian cancer cannot be calculated from available data. Since the rationale for screening is provided by survival data for each FIGO stage, it is reasonable to require evidence of sensitivity for FIGO stage I or at least stage II disease. Furthermore, a screening test for ovarian cancer must fulfil criteria of acceptability and cost-effectiveness which are extremely difficult to define. The incidence of ovarian cancer is relatively low and it is therefore unlikely that any cost benefit will result from ovarian cancer screening.

For the time being, there is no evidence to suggest that the tests or combinations of tests currently available can achieve the combinations of specificity and sensitivity required to screen for early-stage ovarian cancer. There is an urgent need for a randomized controlled trial in order to avoid the introduction of screening for ovarian cancer into clinical practice. To achieve this goal, further preliminary studies are required to identify strategies for improving sensitivity, whilst retaining a high level of sensitivity for early-stage disease using a panel of tumor-associated antigens and to increase specificity using transvaginal color Doppler scanning as a secondary investigation.

METHODS FOR EVALUATION OF ADNEXAL MASSES

The standard evaluation for adnexal masses includes history, physical examination, CA 125 measurements and ultrasound. Although few data are available regarding the accuracy of physical examination to differentiate benign from malignant ovarian masses, it is generally agreed that the clinical impression has little predictive value. We will describe briefly the advantages and disadvantages of all available techniques.

Pelvic findings

Although pelvic findings are of limited value in diagnosis, currently they are the usual means of making a diagnosis. The physician must be alert to the following: (1) a mass in the adnexa; (2) relative immobility due to fixation and adhesions; and (3) irregularity of the tumor, relative insensitivity of the mass, and bilaterality (70% in ovarian cancer versus 5% in benign lesions). Although the ovarian cancer has been described as being hard and knobbly on palpation, close observation challenges this physical finding. Since the ovary often grows at different rates in various parts of the tumor, it is not uncommon to find that areas have grown away from the blood supply. Therefore, there may be areas that are cystic, rubbery, soft, hard and generally variegated.

Pelvic examination is generally regarded as lacking sufficient sensitivity to be of value in the early detection of ovarian cancer. The findings of McFarlane *et al.* are often quoted to support this view[10]. They discovered only six ovarian cancers during a total of 18 753 pelvic examinations performed in 1319 women over a 15-year period (1938–52). It is not possible to reach conclusions concerning either the specificity or sensitivity of pelvic examination from this study.

Some evidence to support the view that vaginal examination lacks sensitivity for the detection of an adnexal mass (either benign or malignant) was provided by the study of Andolf *et al.*[11]. Each patient recruited underwent a pelvic examination prior to ultrasonography. Pelvic examination was reported as normal in 16 benign ovarian cysts, two borderline malignancies and one ovarian carcinoma.

Encouraging levels of specificity were achieved by the combination of CA 125 measurement with either pelvic examination (100%) or ultrasound (99%). This multimodal approach does appear to provide acceptable levels of specificity[12].

Barber and Graber reported an early sign of ovarian cancer which has proved most valuable in diagnosis, that is, the postmenopausal palpable ovary syndrome[13]. It is simply that the palpation of what is interpreted as a normal-sized ovary in the premenopausal woman represents an ovarian tumor in the postmenopausal woman. Patients with postmenopausal palpable ovary syndrome should not be followed or re-evaluated but must be investigated promptly for the presence or absence of an ovarian tumor.

Tumor markers

The anatomical site and structure of the ovaries renders them inaccessible to direct examination but may make ovarian malignancy particularly suitable for detection by a serum marker. Antigen produced by early-stage ovarian cancer may reach the peripheral circulation via lymphatics or blood vessels in the well-vascularized ovarian stroma, whilst antigen shed into the peritoneal cavity may reach the thoracic duct and, hence, venous circulation via diaphragmatic lymphatics.

The most extensively investigated ovarian cancer-associated antigen is CA 125. This antigen is recognized by a murine monoclonal antibody which was produced using an ovarian cancer cell line as an immunogen[14]. Serum levels of CA 125 measured by radioimmunoassay are elevated (>35 U/ml) in over 80% of epithelial ovarian cancers and a smaller proportion of bowel, breast, lung and pancreatic malignancies[15, 16]. This antigen is detectable in the blood up to 5 years before a malignant ovarian tumor is discovered, although it may not be detectable until the tumor has breached its capsule. Unfortunately, other pelvic malignancies also produce elevated values. It is, therefore, not specific for ovarian neoplasms. For example, elevated serum levels have also been reported in a variety of benign pathological and physiological conditions including first trimester pregnancy, endometriosis, pelvic inflammatory disease, pancreatitis and cirrhosis as well as in a proportion of healthy women[17].

The sensitivity of serum CA 125 measurement alone or in combination with other tests for preclinical early-stage ovarian cancer is unknown. An illustrative study has been performed with 1010 postmenopausal volunteers tested at the London Hospital[18]. In this study, one case of ovarian cancer was detected by screening (with a CA 125 level of 32 U/ml) and no others have presented clinically after about 2 years of follow-up. Combined with ultrasound, its sensitivity is still poor – only three pelvic masses were found in 1010 women and only one of these was malignant. The addition of other biochemical markers did not improve the yield.

Ultrasound

Ultrasound examination of the pelvis and abdomen has become the standard diagnostic test for evaluating adnexal masses (Figures 11.1–11.18). Its principal value in this setting involves confirming the mass, differentiating ovarian from uterine or tubal origin, delineating the internal appearance of the mass, and defining any associated abdominal findings. Whether ultrasound can differentiate between benign and malignant pelvic masses has been the subject of many studies. Meire *et al.* first examined the accuracy of ultrasound in delineating a malignant ovarian mass based on size and appearance[19]. In this study, fixed septa, tumor size exceeding 5 cm, and multiloculation were considered ominous for ovarian malignancies. Only 16 of 27 patients with such findings were found to have ovarian cancer. Furthermore, ultrasound has proved disappointing as a means of delineating disease outside the pelvis. Requard *et al.* found that ultrasound could detect only 20% of metastases from the ovary to the rectosigmoid colon, small bowel, and retroperitoneal lymph nodes[20]. Thus, to determine whether the mass is benign or malignant, operative intervention remains necessary.

The accuracy of diagnostic tests used to evaluate an adnexal mass is of great concern to practicing gynecologists. In the preoperative assessment of ovarian masses, the major diagnostic tools are still clinical impression and ultrasound examination. The study performed by Finkler *et al.* has shown that clinical impression and initial ultrasound examination have low overall sensitivities and exceedingly low sensitivities in postmenopausal patients. Even in postmenopausal patients, in whom suspicion of malignancy is greater, the sensitivity of the initial ultrasound examination was found to be 47%[18].

The first systematic approach to the use of pelvic sonography as a screening procedure for early ovarian neoplasms has been done by Campbell and his group[21]. Initially, they showed that ovarian size and morphology as assessed by transabdominal ultrasound examination agreed well with results obtained by direct measurement and observation at laparotomy. Recently, the results of a prospective study of 5479 self-referred women without symptoms have been described in terms of the ovarian masses detected[22], the value of the screening procedure over time, and the development of new screening strategies entailing the use of defined changes in ovarian volume. Five primary ovarian cancers were detected at stage Ia or

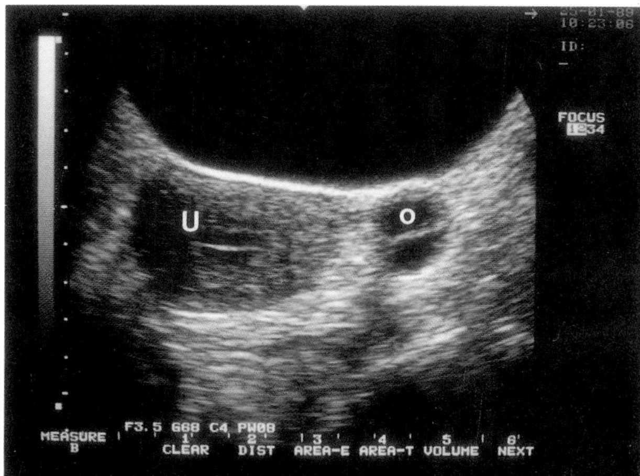

Figure 11.1 Transabdominal scan showing a normal sonogram of the uterus (u) and the left ovary (o). Two mature follicles are observed

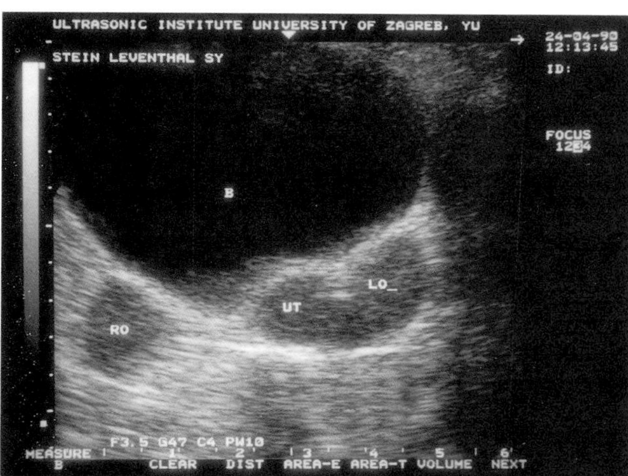

Figure 11.2 Transverse sonogram of a patient with amenorrhea and hirsutism. Both ovaries are enlarged and polycystic. The uterus (UT) is small and the transverse uterine diameter is smaller than the longitudinal ovarian diameter. RO = right ovary; LO = left ovary; B = bladder

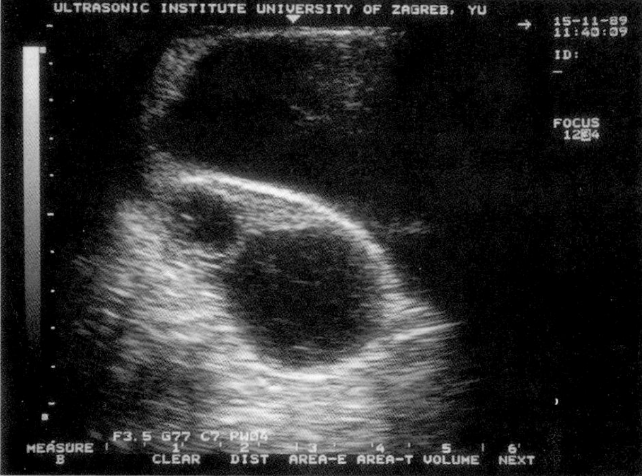

Figure 11.3 An oblique scan of the enlarged ovary. Two separate cystic structures are visualized

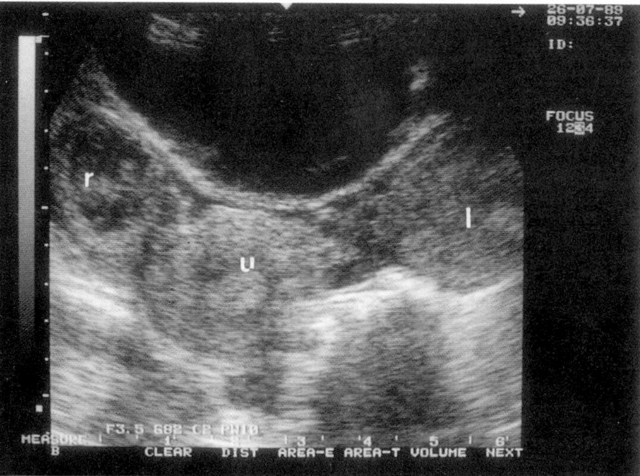

Figure 11.4 Both ovaries are enlarged and predominantly solid structures. r = right ovary; u = uterus; l = left ovary

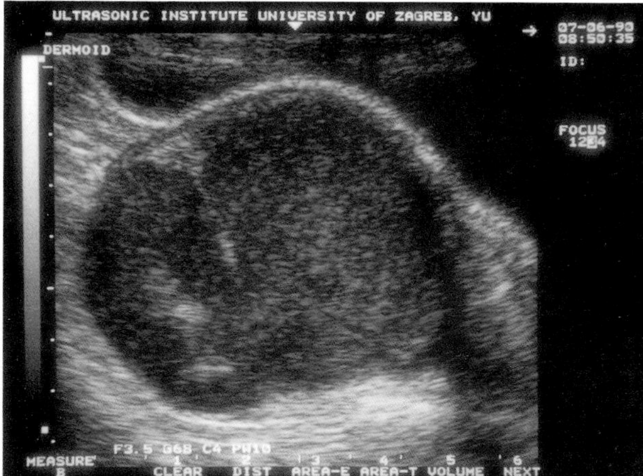

Figure 11.5 A dermoid tumor with a completely solid appearance on the transverse transabdominal sonogram. The dense echo represents teeth, and this is a typical finding of this tumor

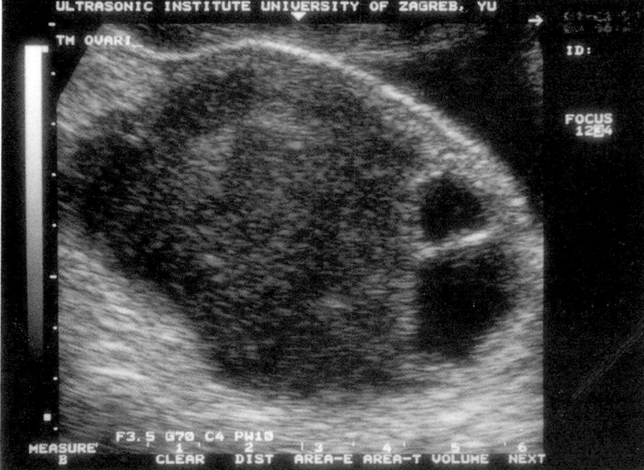

Figure 11.6 A large predominantly solid ovarian mass. Its outline is irregular. A cystic structure with a septa is visualized on the tumor periphery

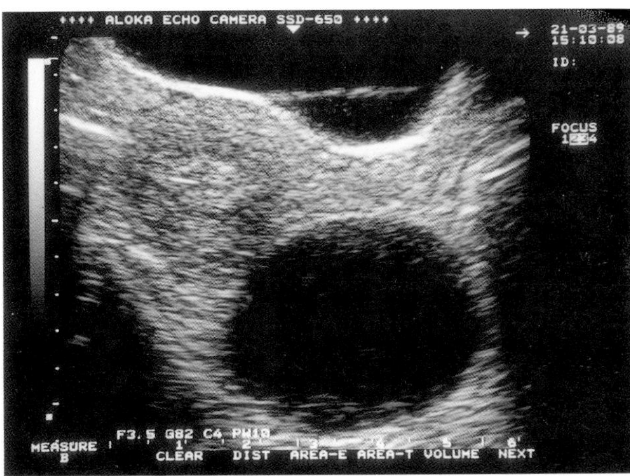

Figure 11.7 A solitary ovarian cyst

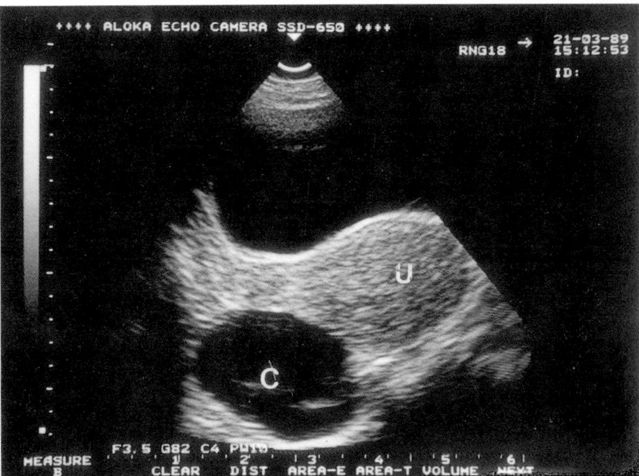

Figure 11.8 Another example of a solitary transonic adnexal cyst (c). u = uterus

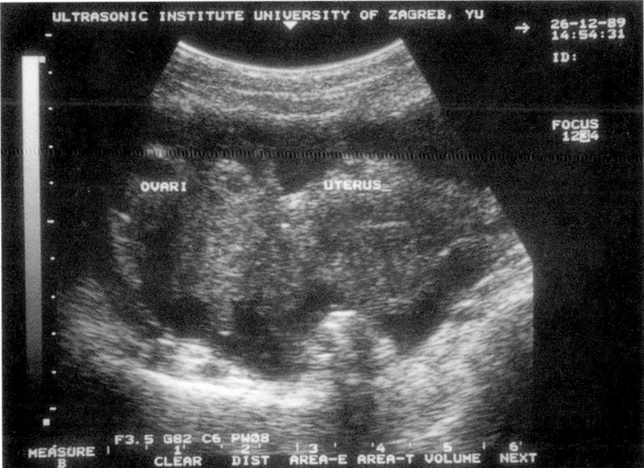

Figure 11.9 A bizarre ultrasound finding of the uterus and the ovary. Ovarian cancer was confirmed by histopathology

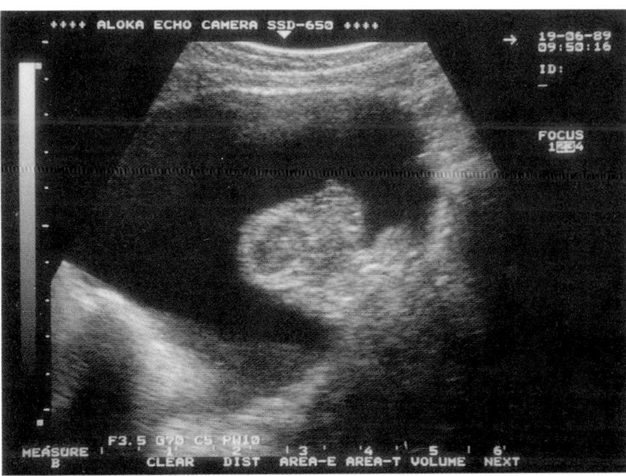

Figure 11.10 A cystic adnexal mass. The strong internal echo represents papillary proliferation. Morphological features are highly suspicious of malignancy

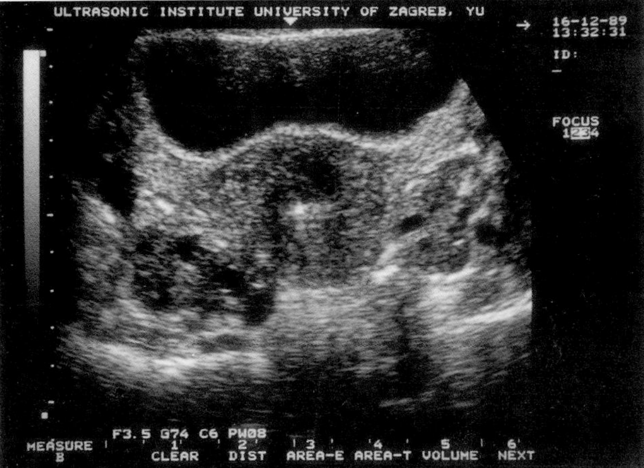

Figure 11.11 A transverse scan of the pelvis. Both ovaries are enlarged, irregularly shaped and extremely unhomogeneous. Ovarian malignancy was suspected

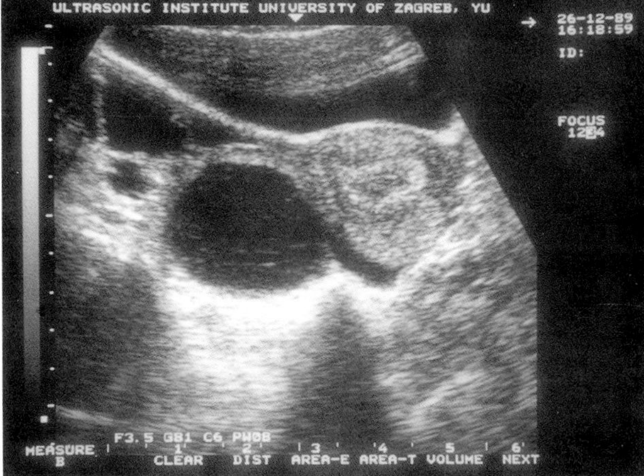

Figure 11.12 A multicystic adnexal mass detected by transabdominal ultrasound

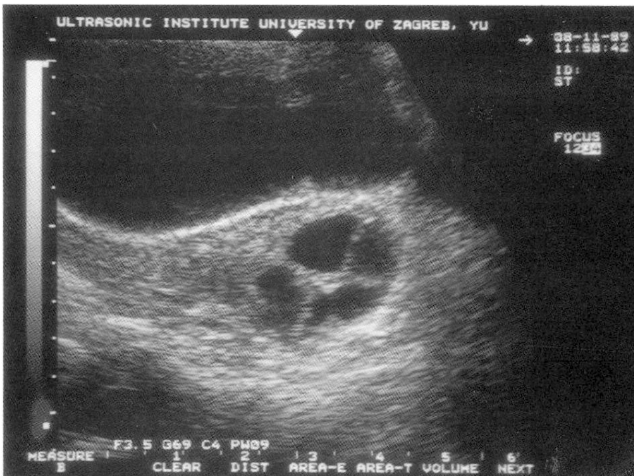

Figure 11.13 A cystic ovarian tumor with well-defined septa on the longitudinal scan

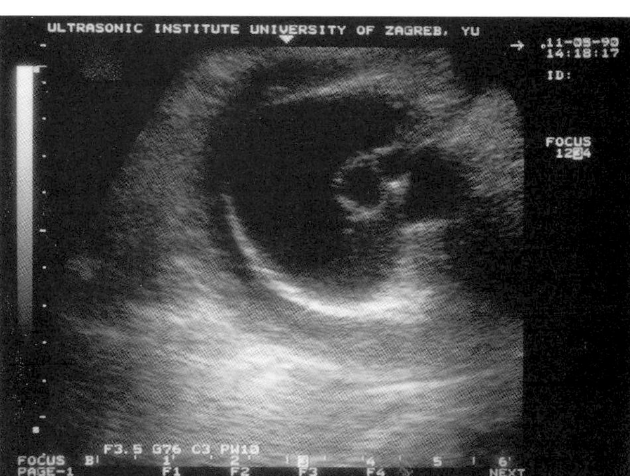

Figure 11.14 A cystadenocarcinoma detected in the initial stage of tumor development. The sonogram is indistinguishable from a benign cystadenoma

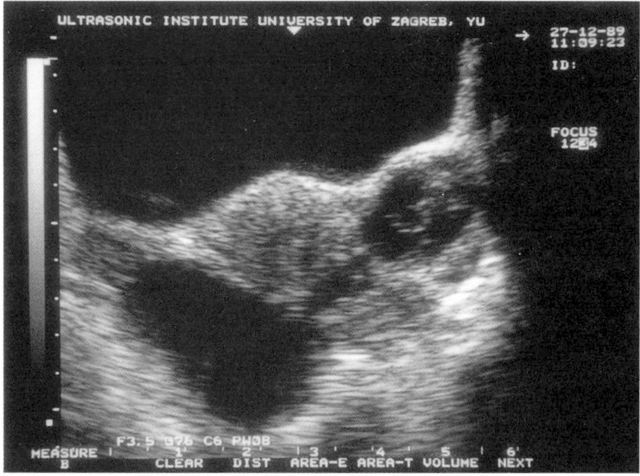

Figure 11.15 Another example of irregular bilateral cystic masses suspected to be malignant adnexal tumors. However, a benign nature (endometriosis) was confirmed by surgery

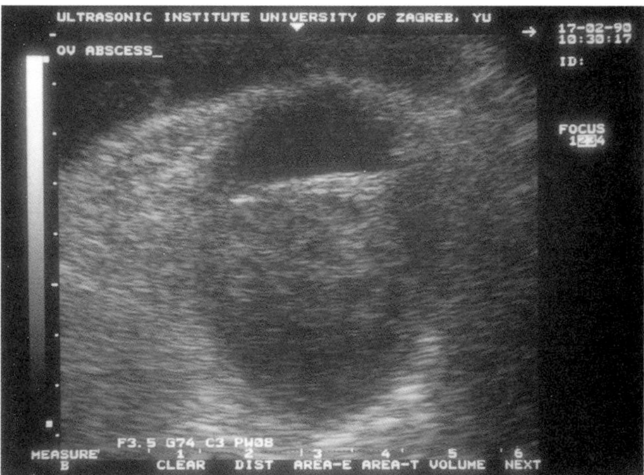

Figure 11.16 A predominantly solid adnexal mass suspected to be an ovarian abscess in a young woman. The final diagnosis was ovarian cancer

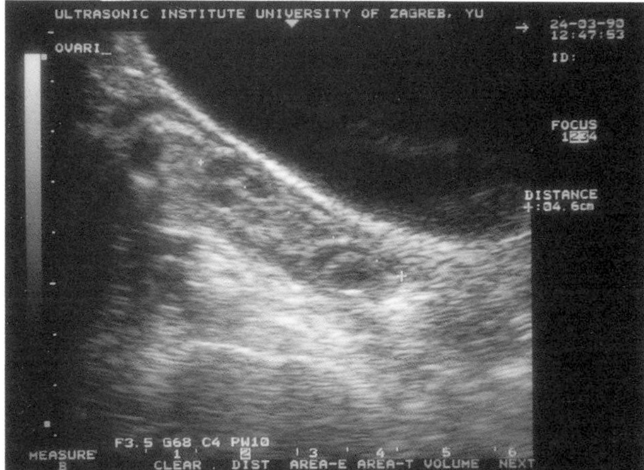

Figure 11.17 A bizarre appearance of ovarian texture, but only a polycystic ovary was diagnosed

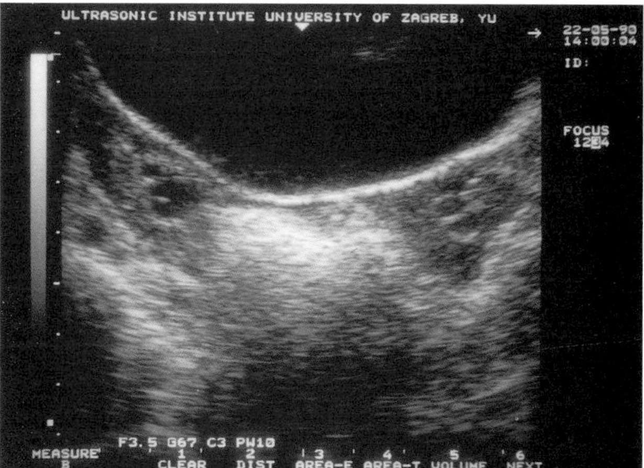

Figure 11.18 Both ovaries are enlarged and polycystic. The uterus was removed 2 years ago

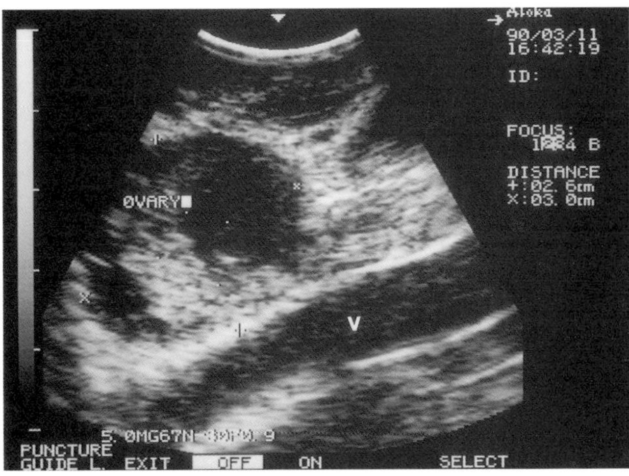

Figure 11.19 A transvaginal scan of the normal-sized left ovary. The internal iliac vessel (v) is also visualized

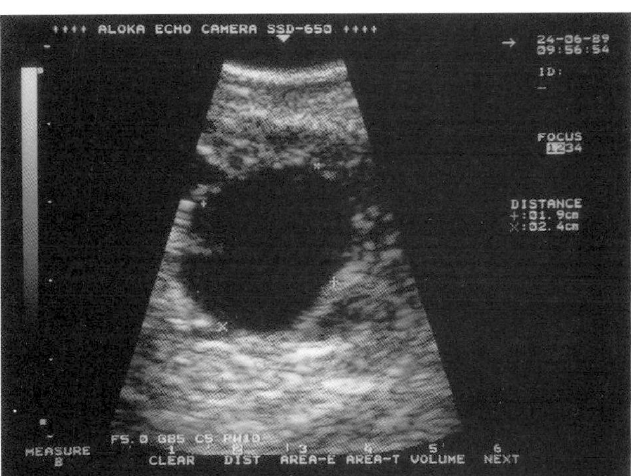

Figure 11.20 A solitary ovarian cyst detected by transvaginal scanning. o = ovarian tissue

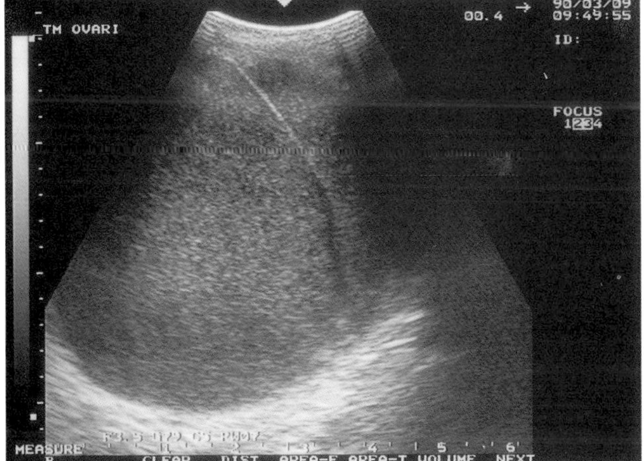

Figure 11.21 Large cystic lobulated adnexal mass visualized by transvaginal ultrasound. A serous cystadenoma was diagnosed

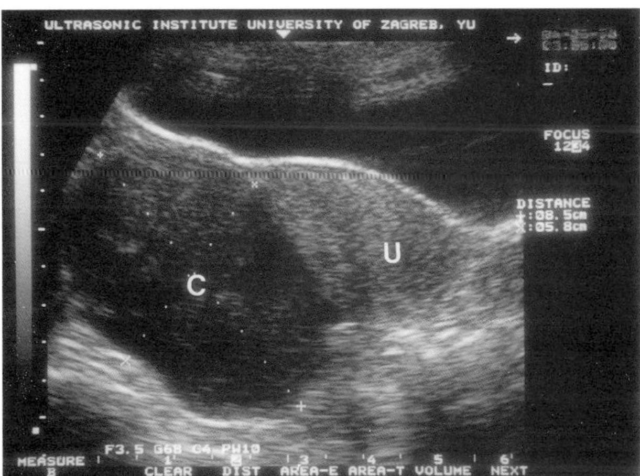

Figure 11.22 This transabdominal scan shows a large transonic cyst (c). u = uterus

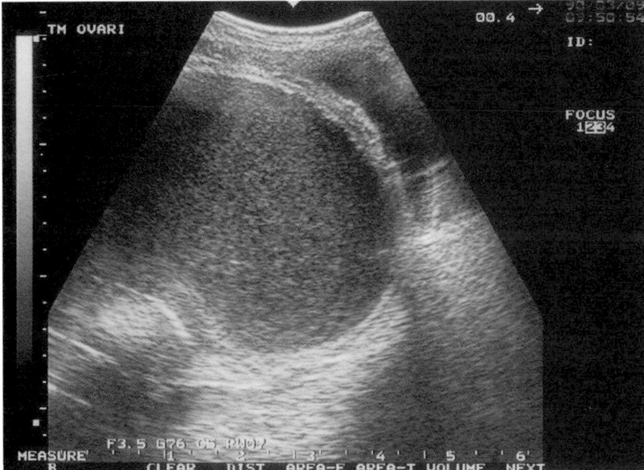

Figure 11.23 The same patient was examined by transvaginal ultrasound. The content of the observed cyst is echogenic. The final diagnosis was mucinous cystadenoma

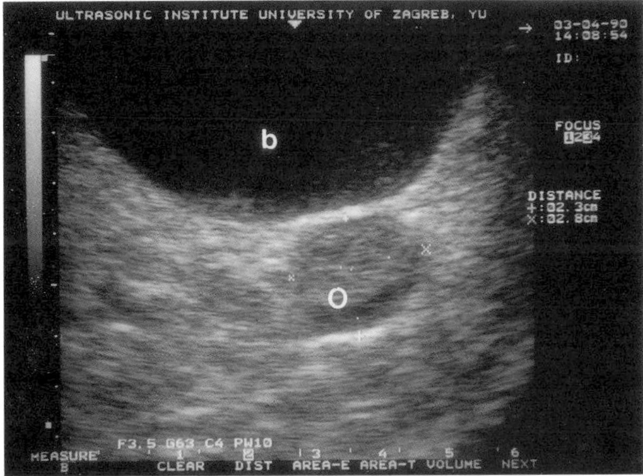

Figure 11.24 A transabdominal scan of the normal postmenopausal ovary (o). b = bladder

Ib, and evidence from a follow-up study at least 1 year after the last screening showed that the detection rate was 100% within the limitation of the study design. A screening procedure based on the presence of abnormal ovarian morphology at the first scan and a defined volume change on rescanning would have given a false-positive rate of 1.6% and a positive predictive value of 2.0% – that is, the odds against a positive screen result indicating the presence of primary ovarian cancer are 1 : 50. This odds ratio is mainly due to the difficulty of distinguishing malignant tumors from benign masses, tumor-like conditions or hydrosalpinges.

The value of routine ultrasound has been examined by Andolf *et al.*[11]. In their study, 805 women attending gynecological outpatient clinics in Sweden were examined. Thirty-nine women had surgery as a result of positive ultrasound findings, including one with malignant and two with borderline ovarian tumors as well as one with cancer of the cecum. Because the results were obtained from symptomatic women, it is difficult to know how to apply the findings to population screening. None of the four tumors detected by ultrasound could be found by digital pelvic examination.

Transvaginal ultrasound

Transvaginal ultrasonography has been used to monitor follicular development[23] and guide needles for oocyte recovery[24]. This approach has two principal advantages. The patient does not need a full bladder and the tip of the probe is nearer to the ovaries (Figures 11.19–11.23). In addition, transabdominal pulsed wave Doppler ultrasound has been used to study blood flow in the ovarian and uterine arteries of healthy and non-treated infertile women[25]. Transvaginal pulsed wave Doppler ultrasound has been used to study blood flow in ovarian arteries in patients being treated with agents to stimulate multiple folliculogenesis[26]. Alternative quantitative methods have been devised and evaluated for the analysis of waveforms[27]. This approach (called Duplex imaging) offers a view of the surrounding anatomy and evidence of blood flow in major vessels, but not in the microcirculation[28]. The technique has been applied to the detection of breast cancer and it has been shown that approximately 90% of localizable tumors exhibit abnormal flow properties (e.g. high peak velocity, increased mean velocity or reverse flow)[29]. More recently, a technique has been developed that shows B-mode gray-scale and color-flow images in the same real-time display. The technique is called color Doppler and offers a quantitative picture of blood flow in the vascular system in relation to the surrounding anatomy[30–34].

It is well known that postmenopausal women have the highest risk for developing ovarian cancer. However, the small size and location of the typical postmenopausal ovary often make sonography difficult (Figures 11.24–11.33). The ovary in the postmenopausal woman averages less than 1.5 cm in size and 2 cm^3 in volume and is frequently located deep in to the bowel[35].

Fleischer performed a study evaluating delineation of normal postmenopausal ovaries in women undergoing hysterectomy and bilateral salpingo-oophorectomy for endometrial carcinoma[36]. He has been able to preoperatively delineate only about 20% of normal ovaries. This indicates possible limits to the accuracy of identifying early tumors as textural abnormalities within normal sized ovaries. Additionally, the sonographic appearance of the tumor does not always correlate with its histological composition[37]. A study of 150 masses in post-menopausal women shows only 3% malignancy in masses less than 5 cm in size[38].

In a very recent study, Bourne *et al.* used transvaginal color Doppler to screen for ovarian cancer[39]. Women were selected on the basis of their medical history and the result of a previous transvaginal ultrasound scan. Thirty women (ten premenopausal (scan taken on days 1–8 of the menstrual cycle) and 20 postmenopausal) had normal ovaries, and 20 had at least one ovary with an abnormal morphology or volume, or both. Two women with a positive result on screening had hydrosalpinges, ten a benign tumor or a tumor-like condition, and eight primary ovarian cancers. No areas of neovascularization were seen in the 30 women with morphologically normal ovaries and the two patients with hydrosalpinges; the pulsatility index ranged from 3.1 to 9.4. Similarly, nine patients (ten affected ovaries with a non-malignant mass) had no signs of neovascularization and the pulsatility index varied from 3.2 to 7.0. One patient with bilateral dermoid cysts containing nests of thyroid-like cells had vascular changes and pulsatility index values of 0.4 and 0.8. Seven patients (eight ovaries) with primary ovarian cancer (one stage IV, four stage III and two stage Ia) showed clear evidence of neovascularization and pulsatility index values from 0.3 to 1.0. One patient with an intraepithelial serous cystadenocarcinoma in a small ovary ($<5\ cm^3$ volume) had no signs of any vascular change and the pulsatility index was 5.5. The authors concluded that transvaginal color flow imaging may be used to identify potentially malignant ovarian masses and help to elucidate the early stages of tumorigenesis.

All described screening programs were in self-selected populations. One which is not selected but which is now almost totally screened by ultrasound is the pregnant population[40].

We believe that detection of angiogenesis will have implications for the treatment of ovarian cysts in pregnancy. Those which are benign can safely be tapped and thus relieve both mother and fetus of the hazards of surgery.

Indeed, on the basis of available evidence, there is sufficient promise in the effectiveness of ovarian cancer screening to set up a randomized trial to see if such screening can reduce mortality from ovarian cancer. Ultrasound has better screening parameters than CA 125 measurements and so a trial of ultrasound has a greater likelihood of success. Undoubtedly, there are sufficient grounds to carry out a trial of ovarian cancer screening provided that some practical problems can be overcome. If such a trial is not carried out soon, screening may be introduced into clinical practice without evaluation and it will then be very difficult, if not impossible, to examine its impact. It is our belief that the early detection of abnormal tumor angiogenesis by transvaginal color Doppler is a new promising diagnostic tool which can be used as a screening test (Figures 11.34–11.50).

ZAGREB EXPERIENCE

The first purpose-built vaginal probe with facilities for color-flow imaging was developed for our use by Aloka Co., Japan. Studies began on 14th November, 1986. Six months later we presented the first paper at the International meeting in Dubrovnik, Yugoslavia. Since that period, more than 15 000 women have passed through our Institute. Most of them were clinically healthy women. However, 1753 women were examined because of clinically and/or ultrasonographically assessed adnexal masses. All patients were examined using the 5 MHz transvaginal probe. The machines employed were Color Doppler SSD-350 and SSD-680 (Aloka Co., Japan). The morphology of the uterus and adnexal region was carefully studied. Color Doppler was used to visualize blood flow in normal pelvic vessels. However, one of the main advantages of color Doppler is the visualization of newly formed tumor vessels. The anatomical position of such vessels depends on the position of the tumor and particularly on its nature. Furthermore, pulsed Doppler was used to quantify such color-coded blood flow. The Pourcelot resistance index was used to assess the impedance of blood flow. At least five separate cardiac cycles were observed and the mean value of the resistance index (RI) was calculated. The mean duration of examination was 20 minutes.

Our own results have shown that it is possible to demonstrate color flow in small vascular branches within tissues in pathological conditions (Figures 11.51–11.59)[31–34, 41–48]. In the observed group of patients with adnexal masses, 680 underwent surgery. Among them, 624 benign adnexal tumors and 56 malignant adnexal tumors were diagnosed by surgery and histopathology (Tables 11.1 and 11.2). The presence of neovascularization was documented preoperatively in almost all malignant ovarian tumors. The characteristic finding was the visualization of small vascular channels within the solid part of the tumor. Being very thin and randomly dispersed within the tissue, such vessels are difficult to find unless transvaginal color Doppler is used. Our non-comparative prospective study has shown that tumor vascularity can be successfully used for the characterization of pelvic tumors (Table 11.3). Sensitivity, specificity and accuracy are acceptably high within the limitation of the study design, and it seems that we should be more concerned about false-negative rather than about false-positive findings. The waveform analysis of signals obtained from the 'hot' color-coded area within masses produced

Table 11.1 Transvaginal color Doppler assessment of benign adnexal mass (n = 624)

		Color flow		
Histopathology	*n*	*Present*	*%*	*RI (2 SD)*
Simple ovarian cyst	270	48	17.8	0.65 (0.10)
Endometriotic cyst	116	17	14.6	0.62 (0.07)
Hydrosalpinx	64	22	34.3	0.70 (0.09)
Dermoid	74	6	8.1	0.64 (0.11)
Pseudocyst	1	1	100	0.39
Serous cystadenoma	53	7	13.2	0.53 (0.06)
Mucinous cystadenoma	46	6	13.0	0.51 (0.08)

Table 11.2 Transvaginal color Doppler assessment of malignant adnexal mass (n = 56)

		Color flow		
Histopathology	*n*	*Present*	*%*	*RI (2 SD)*
Papillary cystadenocarcinoma	4	3	75	0.34 (0.05)
Serous cystadenocarcinoma	21	21	100	0.32 (0.06)
Mucinous cystadenocarcinoma	16	16	100	0.37 (0.02)
Granulosa cell tumor	6	6	100	0.29 (0.09)
Metastatic carcinoma	9	8	89	0.38 (0.03)

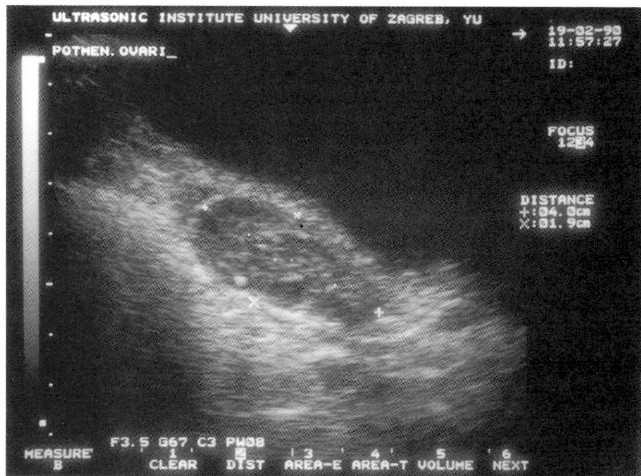

Figure 11.25 An enlarged postmenopausal ovary

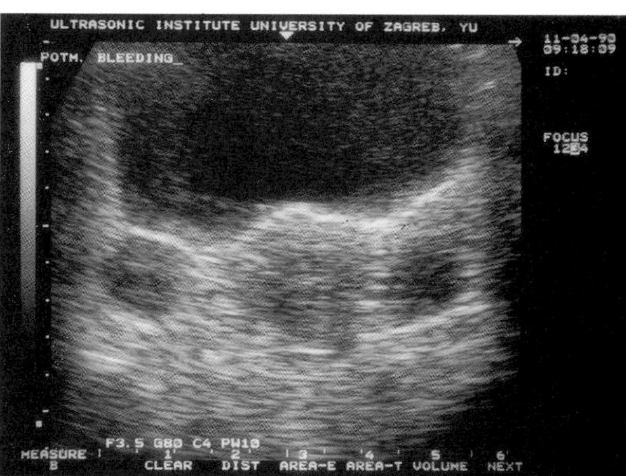

Figure 11.26 The ultrasound finding in a patient with postmenopausal bleeding. Both ovaries are enlarged and hypoechoic

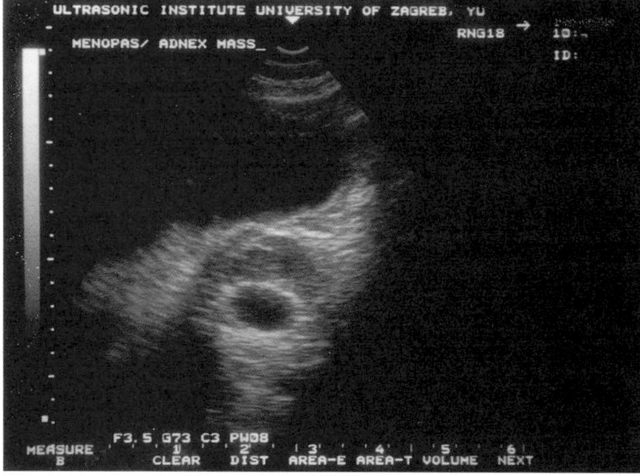

Figure 11.27 A complex adnexal mass detected by transabdominal ultrasound in a postmenopausal woman

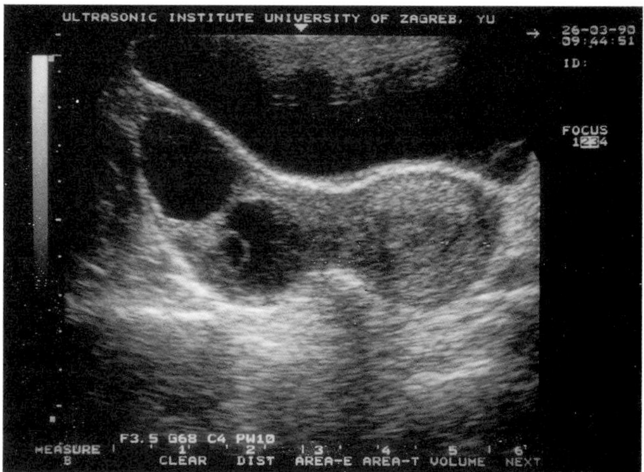

Figure 11.28 A large multilocular cystic mass detected in a postmenopausal woman. Advanced ovarian cancer was proved on histopathology

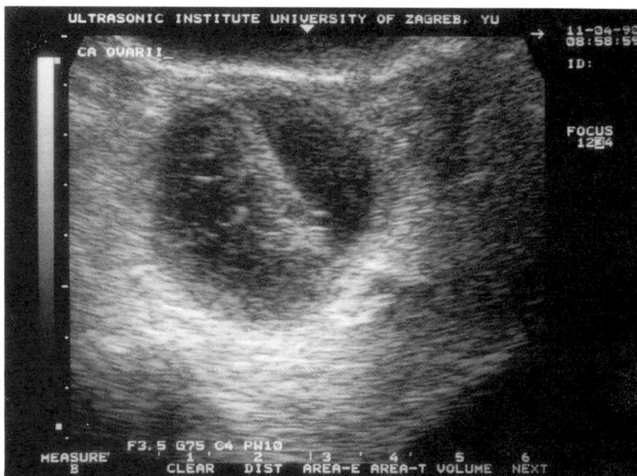

Figure 11.29 A moderately enlarged ovarian mass. Normal ovarian texture could not be visualized

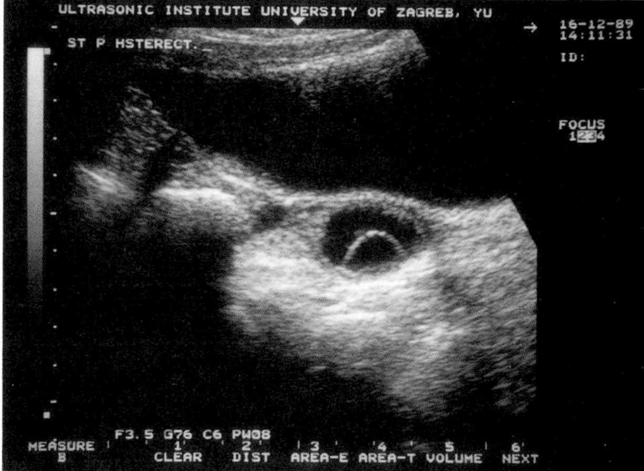

Figure 11.30 A bilocular cystic structure detected in the post-hysterectomy period

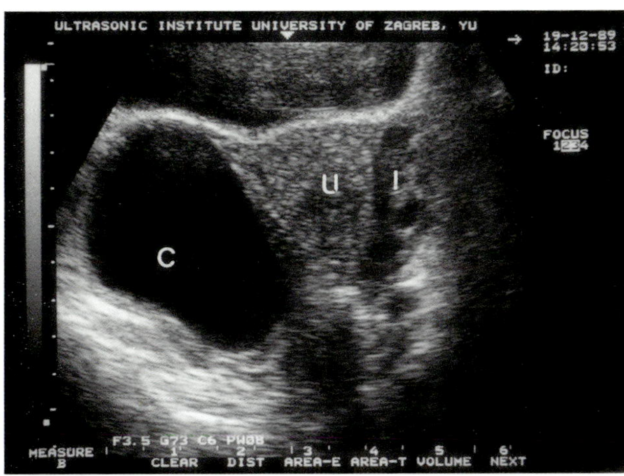

Figure 11.31 A large cyst (c) in a postmenopausal women detected by transabdominal ultrasound. The left ovary (l) is also enlarged. The final diagnosis was cystadenoma. u = uterus

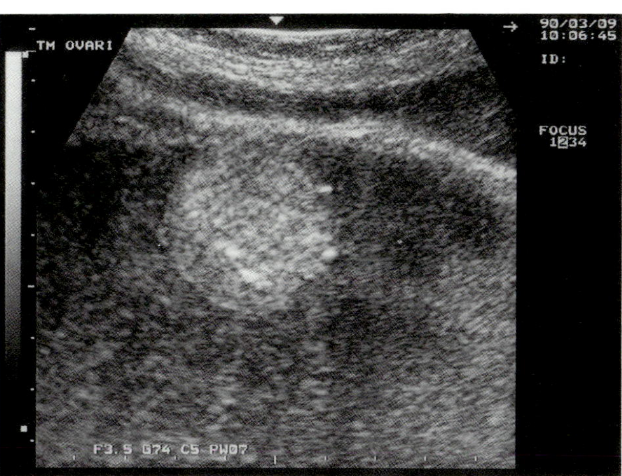

Figure 11.32 Papillary proliferation in the case of a large ovarian tumor detected by transvaginal sonography

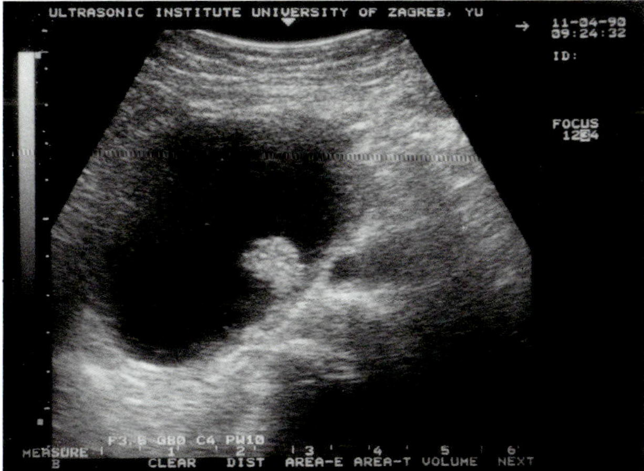

Figure 11.33 Another example of cystic adnexal tumor with papillae

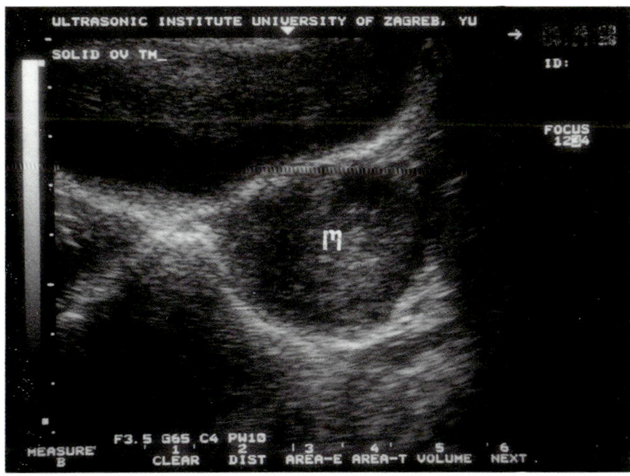

Figure 11.34 Solid ovarian tumor visualized by transabdominal ultrasound. m = solid mass

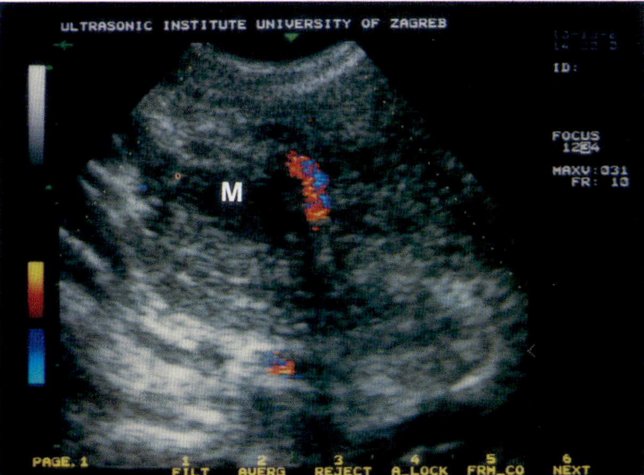

Figure 11.35 The same patient. Transvaginal color Doppler shows tumor blood flow. m = adnexal mass

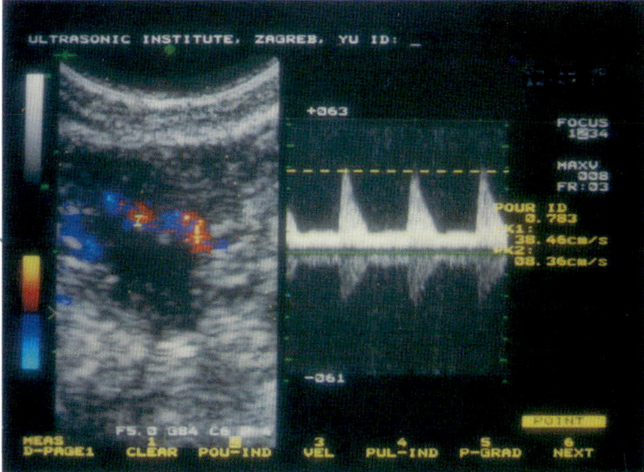

Figure 11.36 The same patient. Pulsed Doppler analysis (right) shows moderate-velocity and high-resistance flow (RI = 0.783). The benign nature of the adnexal tumor was proven by histopathology

Table 11.3 Clinical value of transvaginal color Doppler examination in the preoperative assessment of the tumor nature ($n = 680$)

Color flow and RI < 0.40	*Histopathology* Malignant	Benign	*Total*
Yes	54	1	55
No	2	623	625
Total	56	624	680

Positive predictive value = 98.2%; negative predictive value = 99.7%; sensitivity = 96.4%; specificity = 98.8%; accuracy = 99.5%

excellent results. All malignant tumors had a resistance index lower than 0.40 and all benign tumors, displaying color flow, had a resistance index higher than 0.40. One false-positive finding was obtained in the case of an inflammatory pseudocyst (RI = 0.39). Two false-negative results were noticed. One patient had a papillary cystadenocarcinoma and another metastatic endometrial cancer. Both were postmenopausal women and had highly suspicious morphological findings.

Among apparently healthy women we have diagnosed six primary ovarian malignancies, later on proved by histopathology to be at stage I (Table 11.4). None of the performed examination techniques (clinical examination, transabdominal and transvaginal ultrasound) showed pathological findings. Only transvaginal color and pulsed Doppler presented an abnormal blood flow pattern. Some of these patients were not in the high-risk group for ovarian cancer. They were picked up during the screening program carried out at our Institute.

In an attempt to avoid normal luteal blood flow, we suggest that patients with an adnexal mass suspected to be malignant or those in the screening procedure for ovarian cancer should be examined only during days 3–10 of their menstrual cycle. In the proliferative part of the normal menstrual cycle, it is not possible to visualize any low-impedance high-velocity blood flow. Further examination will, very probably, develop the criteria for accurate differentiation of normal and abnormal blood flow inside ovarian tissue. The velocity analysis in each of the mentioned cases could play a very important role in confirming the final clinical diagnosis.

CONCLUSION

Ovarian cancer, despite being less frequent, kills more women annually than breast and uterine cancer combined. There is no screening test for ovarian

Table 11.4 Some clinical data of apparently healthy patients with abnormal color and pulsed Doppler finding. Primary ovarian cancer at stage I was diagnosed later on ($n = 7$)

Case	*Age* (years)	*Histopathology*	*Stage*	*RI*
1	53	serous cystadenocarcinoma	Ib	0.37
2	59	serous cystadenocarcinoma	Ib	0.36
3	28	mucinous cystadenocarcinoma	Ia	0.35
4	49	papillary cystadenocarcinoma	Ia	—
5	60	mucinous cystadenocarcinoma	Ib	0.40
6	42	serous cystadenocarcinoma	Ia	0.32
7	58	serous cystadenocarcinoma	Ib	0.35

cancer, and most women present with advanced disease. Early detection of ovarian cancer has been an elusive problem. It is generally accepted that annual pelvic examination makes no significant impact on the disease.

Epidemiologic studies have identified a series of associative risk factors in ovarian cancer. These include:

- age, particularly after age 45
- race, whites more susceptible
- higher social class is more at risk
- unmarried, nuns
- infertility
- nulliparity, or first pregnancy delayed after 30 years old
- family history of ovarian cancer
- personal history of neoplasms of breast, colon or endometrium
- presence of Turner's or Peutz–Jeghers syndrome
- high dietary animal fat
- previous chemotherapy
- use of talc or powder on perineum, vagina
- sustained raised levels of GnRH, FSH, LH, e.g. menopause
- adolescent rubella infection

These epidemiologic associations have not been implemented in any formal way to assess their usefulness, but one could be reasonably sure that there would be poor sensitivity and positive predictive values.

Another potential approach could be the utilization of a chemical screen of tumor markers. Some of these that have been associated with ovarian neoplasms include CA 125, α-fetoprotein, carcinoembryonic antigen, monoclonal antibodies, and oncogenes. None of these markers appears to be sensitive enough to be used as a screening tool, although they are being used as a means of monitoring the effectiveness of chemotherapy.

Abdominal ultrasound is useful to identify and delineate the characteristics of cystic ovarian masses. Features of a mass which are of importance are unilocular or multilocular, opaque versus clear fluid, presence of solid components, presence of papillary projections, roughness or smoothness of the internal lining, and the presence of peritoneal fluid. Many of these features can help delineate benign from suspected malignant growths, but once again sensitivities beyond 75% have not been achieved. This figure is too low for a screening test.

Doppler ultrasound identifies the movement of erythrocytes in blood vessels. A signal is obtained and displayed on a monitor. From the display, the contour and frequency return can be calculated. We have learned from Doppler ultrasound that vessels with high resistance–low flow can be differentiated from those with low resistance–high flow. Low-resistance vessels are characteristic of pregnancy-related vessels in the uterus and fetus, the kidney and the brain. Malignant neoplasms may also have vessels with low resistance–high flow. When color is added to the Doppler system, the examiner is able to see many more vessels clearly, particularly small vessels. In most cancers, there is a proliferation of vessels which were formerly identified in radiographic angiology as a tumor 'blush'. Doppler measurements can then be made on these previously inaccessible vessels. We believe that transvaginal color Doppler is a technology which offers the promise of high sensitivity, specificity and positive predictive value for the early detection of ovarian cancer. Transvaginal ultrasound accurately identifies significant morphological components of an ovarian enlargement. Color flow Doppler identifies the presence of abnormal blood vessels, and measurement of the systolic–diastolic components of this flow differentiates vessels into high- and low-velocity states.

Zagreb–New York Screening Program

Recently, a collaborative study on the use of transvaginal color Doppler as a screening test for early detection of ovarian cancer has been undertaken at the Ultrasonic Institute, University of Zagreb (Head, A. Kurjak) and the Department of Obstetrics and Gynecology of the Winthrop University Hospital, Mineola, New York (Head, H. Schulman).

The two institutions undertaking this study are uniquely situated to attract a large number of women for annual routine screening. Large numbers are necessary because of the low incidence of the disease, about 13/10 000 women. At least 5000 women are needed to accomplish a feasibility study, the prelude to a prospective randomized study.

Program goals

The goals of this study are those of a feasibility study, namely to determine via a prospective analysis the incidence of disease detection in a volunteer study population, and the diagnostic accuracy. Equally important is the determination of the consequences of screening, such as intervention rates, subsequent testing, and medical and surgical complications.

Objectives

The first objective is to test the hypothesis that transvaginal ultrasound with color Doppler can detect ovarian cancer in asymptomatic, apparently healthy women. The second objective is to demonstrate that such a screening program will not result in an excessive number of therapeutic interventions.

Data and information required

(1) Basic medical and epidemiologic information, specifically directed to risk factors for ovarian cancer (Table 11.5).
(2) Detailed descriptive information of the ultrasound findings (Table 11.6).
(3) Description of the presence or absence of neovascularization, and Doppler velocimetry indices (Table 11.6).
(4) Follow-up of cases in which interventions are carried out (Table 11.6).
(5) Telephone and letter follow-up of women having apparently normal findings during the period of study.

Population to be studied

Apparently healthy women volunteers, age 40–70, will be offered a screening test. Study entrants should be asymptomatic. Yearly testing will be recommended during this phase of the study. The population in Mineola is expected to be predominantly middle class and white. The population in Zagreb will be predominantly working women, or mothers working at home with children. The pilot program in Zagreb has demonstrated that the threat of this disease is perceived as significant by women included in the program, thereby facilitating the enrolment of an adequate number of subjects. We hope to enrol a minimum of 5000 women in 2–3 years with all women being offered annual examinations.

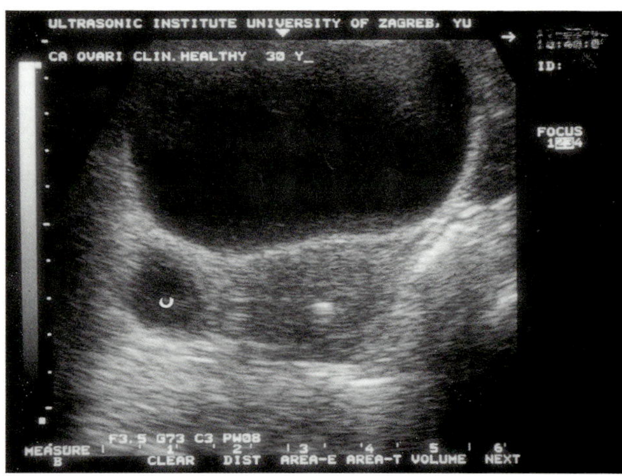

Figure 11.37 The transabdominal sonogram of a 30-year-old clinically healthy woman. Only a slightly enlarged right ovary (o) was noted

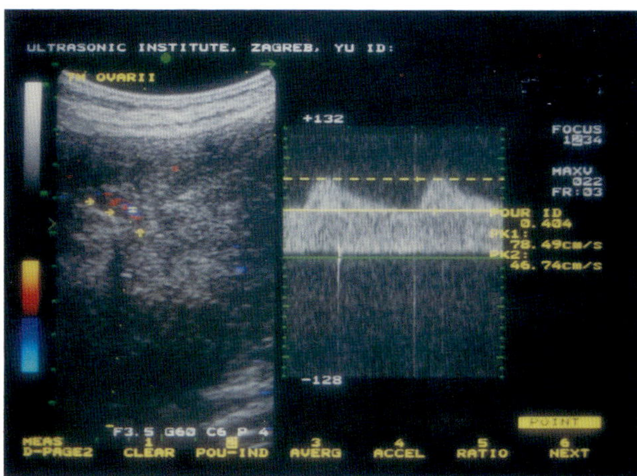

Figure 11.38 The same patient. Transvaginal color shows tumor vessels. Waveform (right) study shows a very high-velocity and very low-resistance blood flow. Ovarian cancer at stage Ia was confirmed on histopathology

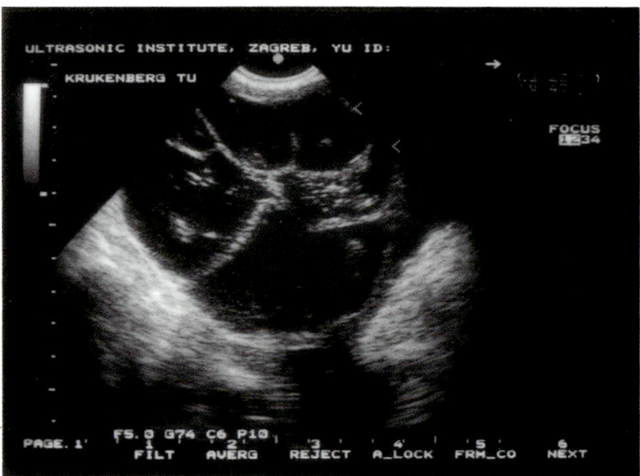

Figure 11.39 Transabdominal scan of the adnexal mass suspected to be Krugenberg's tumor. Tumor septa can be clearly visualized

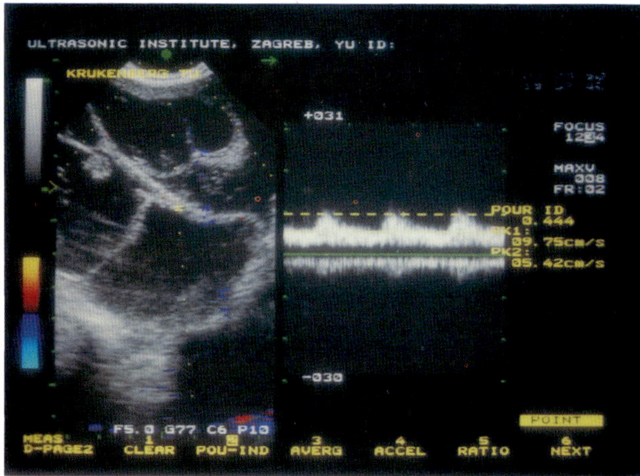

Figure 11.40 The same patient. Blood flow is detected in the septal part by transvaginal color Doppler. Pulsed Doppler (right) shows low-velocity and low-resistance flow

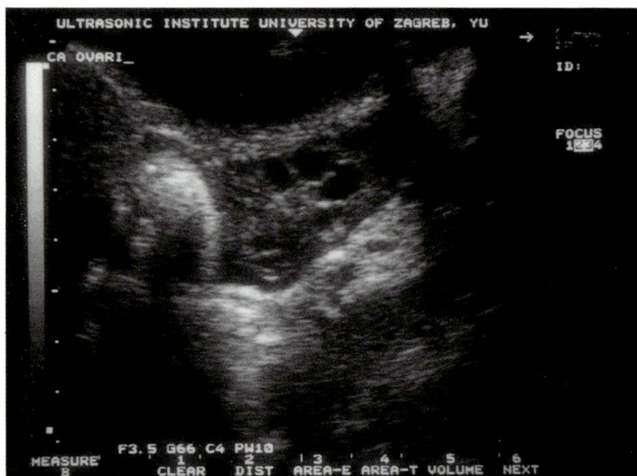

Figure 11.41 A transabdominal scan of the enlarged predominantly solid ovary

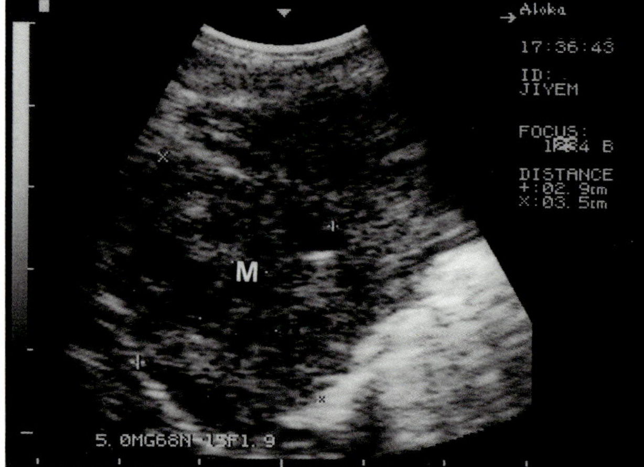

Figure 11.42 The same patient. Transvaginal ultrasound shows only a slightly enlarged ovary. m = ovarian mass

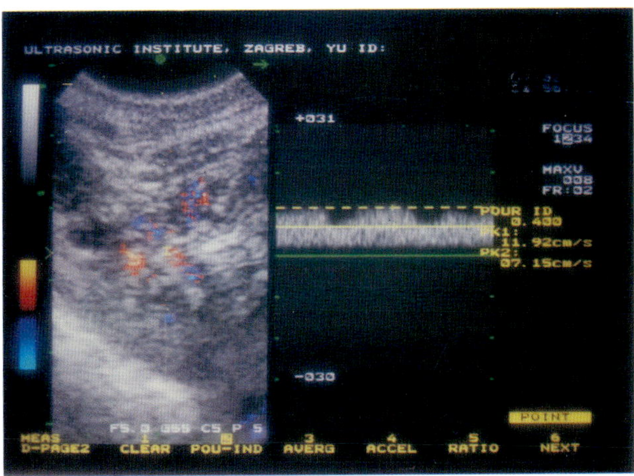

Figure 11.43 The same patient. Abundant color flow is visualized by transvaginal color Doppler. Pulsed Doppler (right) shows a very low resistance of blood flow. A malignant ovarian tumor was diagnosed

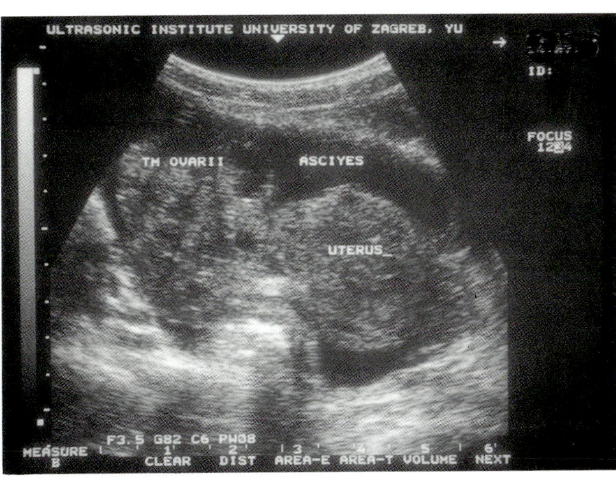

Figure 11.44 An example of an ovarian tumor. Transabdominal ultrasound supports the diagnosis of malignancy

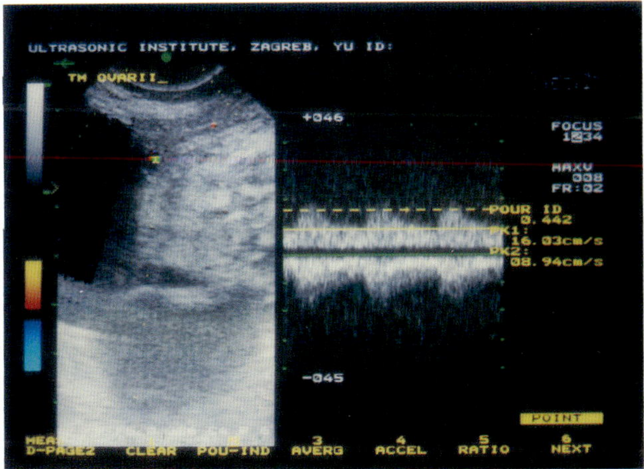

Figure 11.45 The same patient. Transvaginal color Doppler presents tumor neovascularization

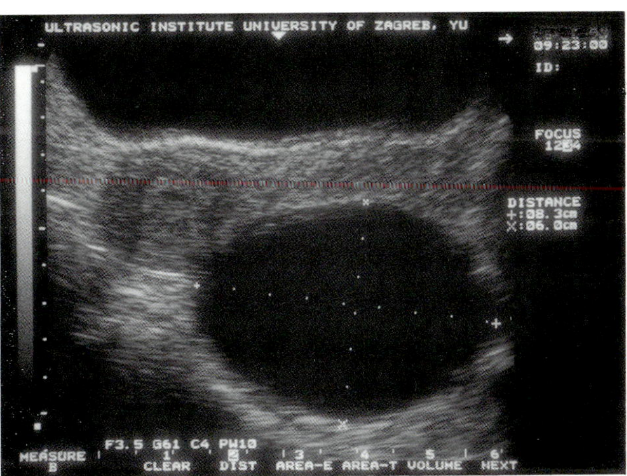

Figure 11.46 The transabdominal scan. A large solitary ovarian cyst is visible

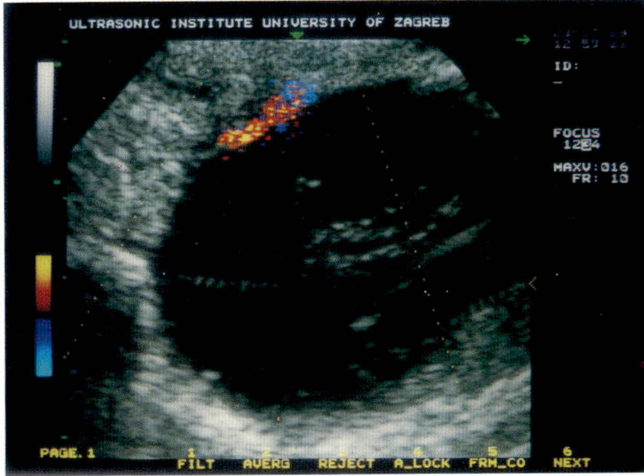

Figure 11.47 Transvaginal color Doppler. Blood flow was detected on the periphery of the cyst

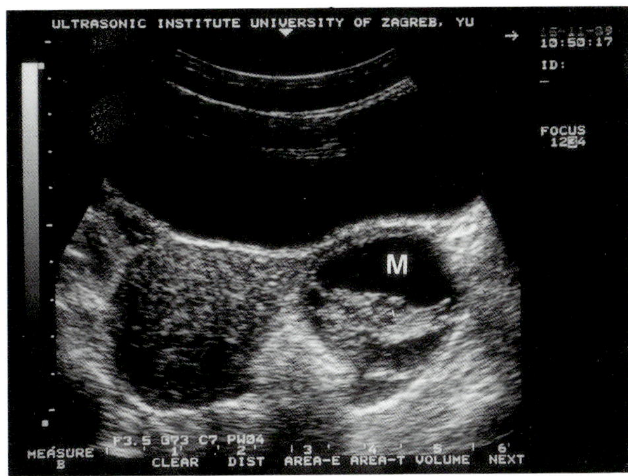

Figure 11.48 Transabdominal scan of a complex adnexal mass (M)

Table 11.5 Ovarian screening program of Zagreb and New York. A questionnaire

(Please answer 'yes' or 'no' unless specific information is requested)

Name ______
Address ______
Telephone ______
Age ______
Race ______
Education in years ______
Occupation ______
Married, previously married ______
Number of pregnancies ______
Number of abortions, miscarriages ______
Number of children ______
Age of oldest child ______
Personal history of cancer, type ______
Did you have Turner's or Peutz–Jeghers syndrome? ______
Have you had chemotherapy? ______
Do you use powders or sprays around the vagina after bathing?
______

Have you menopaused? ______
Have you had German measles? ______
Have you ever had surgery on female organs? Type? ______

Medications taken regularly ______
Contraceptives used ______
Family history cancer, type ______
Last pelvic examination ______
Previous ultrasound, X-ray pelvis: data ______
______

Table 11.6 Ovarian tumor ultrasound–Doppler classification (circle all characteristics seen and add for a score)

Patient name ______ Date ______ Institution ______

	Fluid	*Internal borders*	*Size*
Unilocular	clear (0)	smooth (0)	
	internal echoes (1)	irregular (2)	
Multilocular	clear (1)	smooth (1)	
	internal echoes (1)	irregular (2)	
Cystic-solid	clear (1)	smooth (1)	
	internal echoes (2)	irregular (2)	
Papillary projections	suspicious (1)	definite (2)	
Solid	homogeneous (1)	echogenic (2)	
Peritoneal fluid	absent (0)	present (1)	

Color Doppler	*RI index*	*Velocity*
No vessels seen (0)	(0)	
Regular separate vessels (1)	> 0.40 (1)	
Randomly dispersed vessels (2)	< 0.41 (2)	

If suspected corpus luteum, repeat in next menstrual cycle in proliferative phase

Score	*Ultrasound*	*Color*
≤ 2	benign	benign
3–4	questionable	questionable
>4	suspicious	

Outcome
Spontaneous resolution ______
Surgery, type ______
Pathologic diagnosis ______
Complications ______
Follow up, date and status ______
Length of hospital stay ______

Experimental or survey procedures

The experimental design for both sites is similar. In Mineola, a broad-based announcement of the availability of this screening test for ovarian cancer will be made in the community. Testing will be free. Mineola is the central train station site for the Long Island Railroad going into Manhattan. Hundreds of women pass by the outpatient facility daily, and the plan is to structure services so that these women can avail themselves of the service before and after work. Businesses, hospitals and women's groups will also be contacted. In Zagreb, workers have been sent to the Ultrasonic Institute for testing.

Each subject will fill out a questionnaire and then undergo the examination. Transvaginal ultrasound is generally done in the lithotomy position. The subject need not undress but merely remove her underwear. The urinary bladder should be emptied. The vaginal probe is covered by a sterile condom containing contact gel. The probe diameter is 3 cm and therefore the examination should be painless. It may be inserted into the vagina until there is optimal visualization. This usually occurs at 4–5 cm depth. The testing should average 10–15 minutes with a maximum time of 25 minutes.

The examinations will be carried out by physicians only during the first 6 months. After that, medical technicians or nurses may be used after an adequate training period.

All examinations will require hard-copy photographic documentation.

Data collection methods

Each subject will fill out the questionnaire. The physician ultrasonographer will review the data. The examining sonographer will be responsible for obtaining a hard-copy photograph of the ovaries, and for filling out the data required. The data sheet and medical information will be reviewed by each of the senior investigators at their respective institutions. The data will be placed into a computer program created at Winthrop. Software used for creating these files is Filemaker 4 created by Nashoba systems and now marketed by Claris. Most data items are self-explanatory except for the appearance of color flow and the calculation of the resistance index. Color flow is considered to be absent if there is no significant pattern of vessels; the ones with the brightest color are selected for Doppler insonation and measurement of systolic–diastolic frequencies. The data forms are shown in Tables 11.5 and 11.6.

Data analysis

Standard validity analyses will be performed. These will include determination of *sensitivity*, *specificity*, and *positive* and *negative predictive values*. A true positive will be defined as a woman having a malignant neoplasm of the ovary. Calculations will also be made for the sensitivity of all neoplasms, e.g. benign and malignant. Functional growths such as simple cysts, corpus luteum or follicular cysts in which interventions are carried out will be considered as false positives.

The *reliability* of the test will be evaluated in those with suspicious or positive findings. All women will be offered a second examination by the primary sonographer and a second opinion investigator. This will allow calculation of the inter- and intraobserver error. The time frame should be limited to no more than a 2-week interval between examinations. Reliability tests will be calculated from the ultrasound score and the Doppler indices.

The *yield* of the screening program will be defined as the amount of previously unrecognized disease that is diagnosed and brought to treatment as a result of screening. This will be expressed as a ratio per 1000 women.

The *prevalence* of the disease in this population will be determined. Previous data suggest a wide variation in different regions, and this may impact on the sensitivity of the test.

REFERENCES

1. Dykes, P.W., Bradwell, A.R., Chapman, C.E. and Vaugham, A.T.M. (1987). Radioimmunotherapy of cancer: clinical studies and limiting factors. *Cancer Treat. Rev.*, **14**, 87
2. Sands, H. and Jones, P.K. (1990). Physiology of monoclonal antibody accretion by tumors. In Goldenberg, J. (ed.) *Cancer Imaging with Radioactive Antibodies*. (Boston: Martinus Nijhoff) in press
3. Jain, R.K., Weissbrod, J. and Wei, J. (1980). Mass transfer in tumors: characterization and applications in chemotherapy. *Adv. Cancer Res.*, **33**, 251
4. Gerlowski, L.E. and Jain, R.K. (1983). Physiologically-based pharmacokinetics: principles and applications. *J. Pharm. Sci.*, **72**, 1103
5. Barber, H.R.K. (1984). Ovarian cancer: diagnosis and management. *Am. J. Obstet. Gynecol.*, **150**, 910
6. Cuckle, H.S. and Wald, N.J. (1984). Principles of screening. In Wald, N.J. (ed.) *Antenatal and Neonatal Screening*, pp. 125–39. (Oxford: Oxford University Press)
7. Cuckle, H.S. and Wald, N.J. (1989). The evaluation of screening tests for ovarian cancer. In Sharp, F., Mason, W.P. and Leake, R.E. (eds.) *Ovarian Cancer*, p. 229. (London: Chapman and Hall Medical)
8. Campbell, S., Collins, W.P., Royston, P., Bourne, T.H., Bhan, V. and Whitehead, M.I. (1990). Developments in ultrasound screening for early ovarian cancer. In Sharp, F., Mason, W.P. and Leake, R.E. (eds.) *Ovarian Cancer*, p. 217. (London: Chapman and Hall Medical)
9. Jacobs, I.J. and Orham, D.A. (1989). Potential screening tests for ovarian cancer. In Sharp, F., Mason, W.P. and Leake, R.E. (eds.) *Ovarian Cancer*, p. 197. (London: Chapman and Hall Medical)
10. McFarlane, C., Strugis, M.C. and Fetterman, F.S. (1955). Results of an experiment in the control of cancer of the female pelvis organ and report of a fifteen year research. *Am. J. Obstet. Gynecol.*, **69**, 294
11. Andolf, E., Svalenius, E. and Astedt, B. (1986). Ultrasonography for early detection of ovarian carcinoma. *Br. J. Obstet. Gynaecol.*, **93**, 1286
12. Jacobs, I.J., Stabile, I., Bridges, J. *et al.* (1988). Multimodal approach to screening for ovarian cancer. *Lancet*, **1**, 268
13. Barber, H.R.K. and Graber, E.A. (1971). The PMPO syndrome (postmenopausal palpable ovary syndrome). *Obstet. Gynecol.*, **38**, 921
14. Best, R.C., Feeney, M., Lazarus, H. *et al.* (1981). Reactivity of a monoclonal antibody with human ovarian carcinoma. *J. Clin. Invest.*, **68**, 1331
15. Bast, R.C., Klug, T.L., St. John, E. *et al.* (1983). A radioimmunoassay using a monoclonal antibody to monitor the course of epithelial ovarian cancer. *N. Engl. J. Med.*, **309**, 169
16. Haga, Y., Sakamoto, K., Egami, H. *et al.* (1986). Clinical significance of serum CA 125 values in patients with cancers of the digestive system. *Am. J. Med. Sci.*, **292**, 30
17. Jacobs, I.J. and Bast, R.C. (1989). The CA 125 tumor associated antigen; a review of the literature. *Hum. Reprod.*, **4**, 1
18. Finkler, N.J., Benacerraf, B., Lavin, P.T. *et al.* (1988). Comparison of serum CA 125, clinical impression and ultrasound in the preoperative evaluation of ovarian masses. *Obstet. Gynecol.*, **72**, 659
19. Meire, H.B., Farrant, P. and Guha, T. (1978). Distinction of benign from malignant ovarian cysts by ultrasound. *Br. J. Obstet. Gynaecol.*, **85**, 893
20. Requard, C.K., Mettler, F.A. and Wicks, J.D. (1981). Preoperative sonography of malignant ovarian neoplasms. *Radiology*, **137**, 79
21. Campbell, S., Goessens, L., Goswamy, R. and Whitehead, M.I. (1982). Real-time ultrasonography for the determination of ovarian morphology and volume. A possible early screening test for ovarian cancer. *Lancet*, **1**, 425
22. Bhan, V., Amso, N., Whitehead, M.I. *et al.* (1989). Characteristics of persistent ovarian masses in asymptomatic women. *Br. J. Obstet. Gynaecol.*, **96**, 1384
23. Gonzales, C.J., Curson, R. and Parsons, J. (1988). Transabdominal versus transvaginal ultrasound scanning of ovarian follicles: are they comparable? *Fertil. Steril.*, **50**, 657

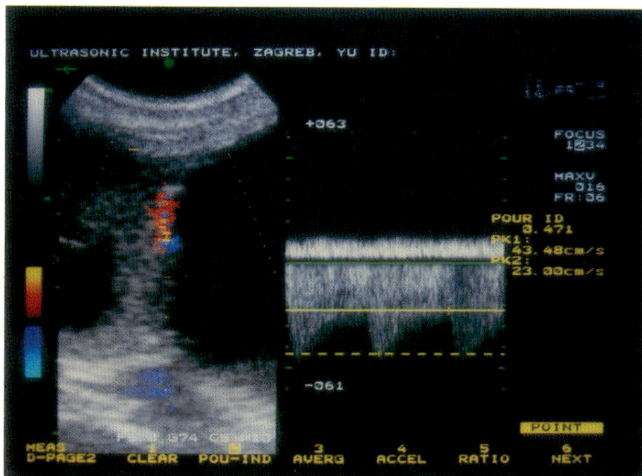

Figure 11.49 Blood flow was detected in the solid part of the same tumor. This is a borderline Doppler finding

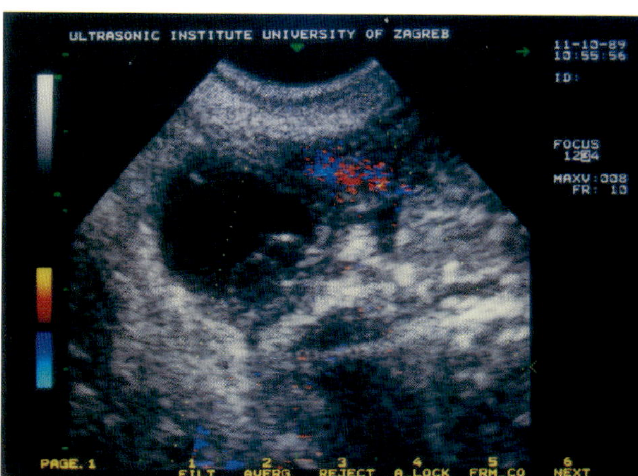

Figure 11.50 Another example of tumor vessels visualized by color Doppler

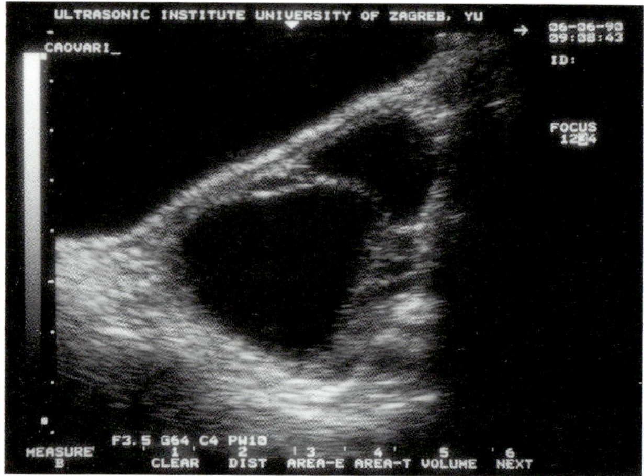

Figure 11.51 A transabdominal sonogram of suspected ovarian cancer

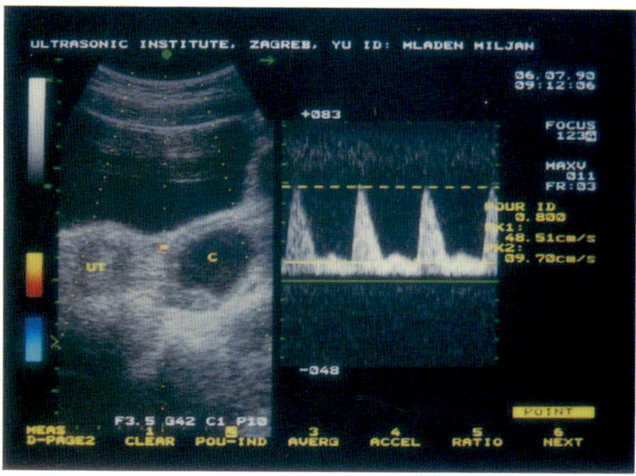

Figure 11.52 The same patient. Color Doppler shows blood flow on the tumor periphery. Pulsed Doppler (right) indicates a normal finding. The benign nature of the tumor (endometriosis) was confirmed by histopathology. UT = uterus; C = ovarian tumor

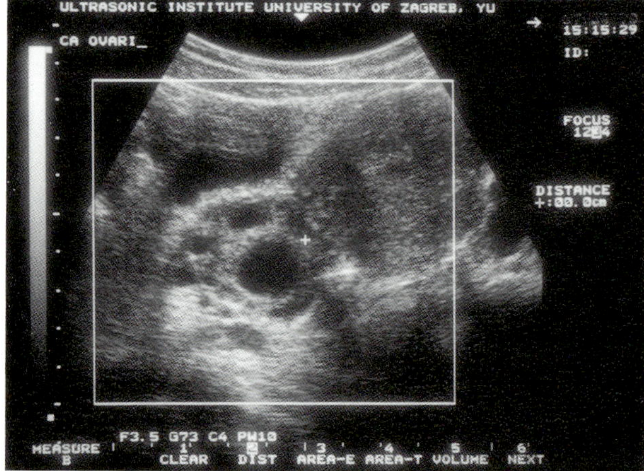

Figure 11.53 Another example of a complex ovarian mass

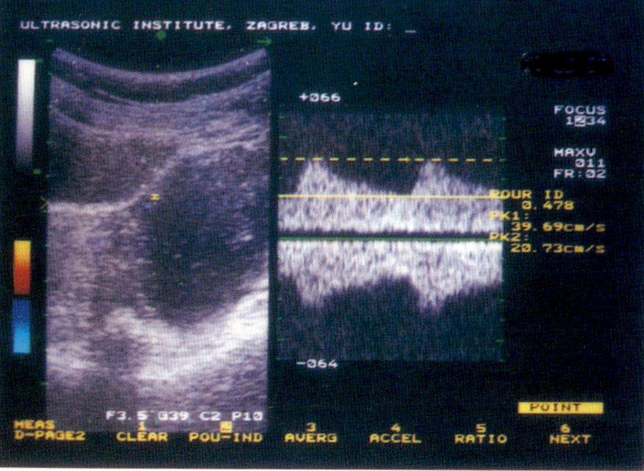

Figure 11.54 Doppler finding shows moderate-velocity and low-resistance flow. Borderline case

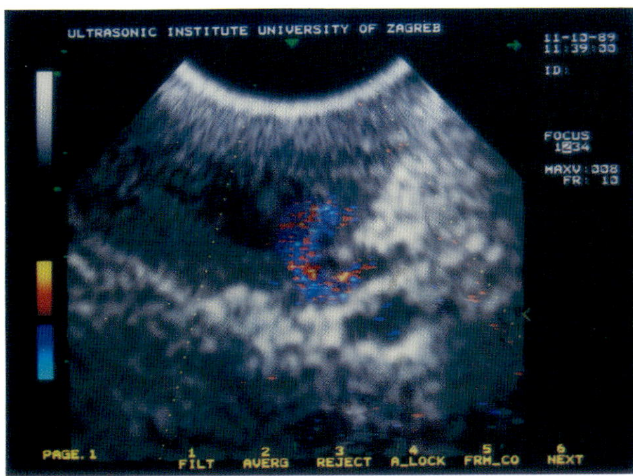

Figure 11.55 The tumor vasculature visualized by transvaginal color Doppler

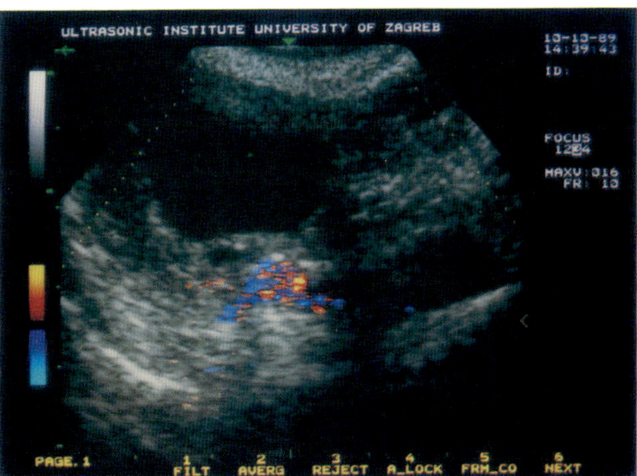

Figure 11.56 Another example of neovascularization

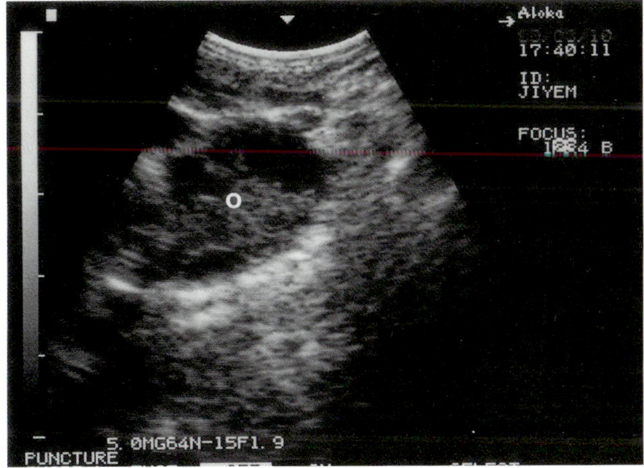

Figure 11.57 Transvaginal scan of the ovary (o) in the luteal part of the cycle

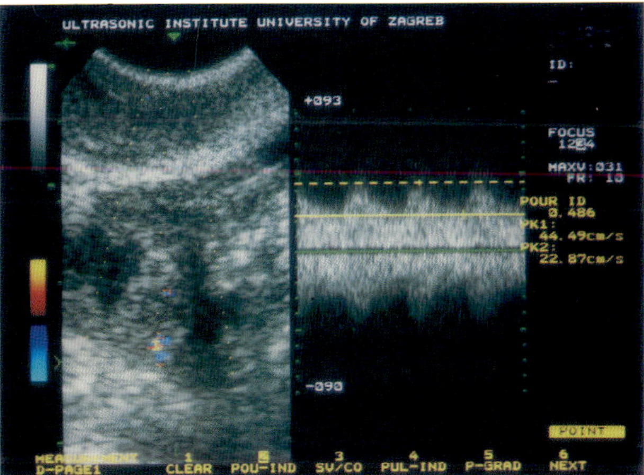

Figure 11.58 Color and pulsed Doppler showing neovascularization of corpus luteum. Such a finding should not be mistaken for tumor angiogenesis

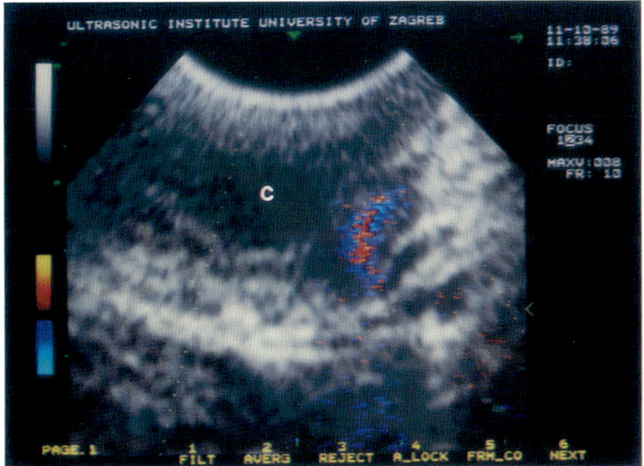

Figure 11.59 Another example of corpus luteum vascularization. c = corpus luteum

24. Feichtinger, W. and Kemeter, P. (1986). Transvaginal sector scan sonography for needle guided transvaginal follicle aspiration and other applications in gynecologic routine and research. *Fertil. Steril.*, **45**, 722
25. Kurjak, A. and Jurković, D. (1986). New ultrasonic technique for assessing circulation in female pelvis. In Kurjak, A. and Kossoff, G. (eds.) *Recent Advances in Ultrasound Diagnosis*, Vol.5, p.129 (Amsterdam, New York, Oxford: Excerpta Medica)
26. Deutinger, J., Reinthaller, A. and Bernaschek, G. (1989). Transvaginal pulsed Doppler measurement of blood flow velocity in the ovarian arteries during cycle stimulation and after follicle puncture. *Fertil. Steril.*, **51**, 466
27. Thompson, R.S., Trudinger, B.J. and Cook, C.M. (1988). Doppler ultrasound waveform indices: A/B ratio, pulsatility index and Pourcelot ratio. *Br. J. Obstet. Gynaecol.*, **95**, 581
28. Thaler, I., Manor, D., Rottem, S., Timor-Tritsch, I.E., Brandes, J.M. and Itskovitz, J. (1990). Hemodynamic evaluation of female pelvic vessels using a high-frequency transvaginal image-directed Doppler system. *J. Clin. Ultrasound*, **18**, 364
29. Minasian, H. and Bamber, J.C. (1982). A preliminary assessment of an ultrasonic Doppler method for the study of blood flow in human breast cancer. *Ultrasound Med. Biol.*, **8**, 357
30. Rubin, J.M., Carson, P.L., Zlotecki, R.A. and Ensminger, W.D. (1987). Visualization of tumor vascularity in a rabbit VX2 carcinoma by Doppler flow mapping. *J. Ultrasound Med.*, **6**, 113
31. Kurjak, A., Jurković, D., Alfirevic, Z. and Žalud, I. (1990). Transvaginal color Doppler imaging. *J. Clin. Ultrasound*, **18**, 227
32. Kurjak, A., Žalud, I., Jurković, D., Alfirevic, Z. and Miljan, M. (1989). Transvaginal color Doppler for the assessment of pelvic circulation. *Acta Obstet. Gynecol. Scand.*, **68**, 131
33. Kurjak, A. (1989). Transvaginal color Doppler in the assessment of pelvic circulation. *Jpn. J. Med. Ultrasound*, **16** (Suppl. II), 1
34. Kurjak, A. and Jurković, D. (1989). Transvaginal color Doppler in the assessment of pelvic masses. Proceedings of *6th World Congress on In Vitro Fertilization and Alternate Assisted Reproduction*, Jerusalem, Abstr. p.26
35. Grandberg, S. and Wikland, M. (1987). Comparison between endovaginal and transabdominal transducers for measuring ovarian volume. *J. Ultrasound Med.*, **6**, 649
36. Fleischer, A.C. (1988). Transvaginal sonography helps find ovarian cancer. *Diagn. Imaging*, **10**, 124
37. Moyle, J.W., Rochester, D., Sider, L. *et al.* (1983). Sonography of ovarian tumors: predictability of tumor type. *Am. J. Roentgenol.*, **141**, 985
38. Rulin, M.C. and Preston, A.L. (1987). Adnexal masses in postmenopausal women. *Obstet. Gynecol.*, **70**, 578
39. Bourne, T., Campbell, S., Steer, C., Whitehead, M.I. and Collins, W.P. (1989). Transvaginal colour flow imaging: a possible new screening technique for ovarian cancer. *Br. Med. J.*, **299**, 1367
40. Editorial (1990). First catch your deer. *Lancet*, **336**, 147
41. Kurjak, A., Žalud, I., Alfirevic, Z. and Jurković, D. (1990). The assessment of abnormal pelvic blood flow by transvaginal color Doppler. *Ultrasound Med. Biol.*, in press
42. Kurjak, A. and Žalud, I. (1990). Transvaginal color Doppler in the characterization of pelvic masses. *Ultraschall Med.*, in press
43. Kurjak, A. and Žalud, I. (1990). Transvaginal color Doppler. In Chervenak, F.A., Isaacson, G. and Campbell, S. (eds.) *Textbook of Ultrasound in Obstetrics and Gynecology*. (Boston: Little, Brown & Co.) in press
44. Kurjak, A. (1990). Transvaginal color Doppler in the detection of ovarian malignancy. The *7th Congress of the European Federation of Societies for Ultrasound in Medicine and Biology*, Jerusalem, May 6–11, Abstr. p. 91
45. Kurjak, A., Žalud, I. and Grljusic, V. (1990). Transvaginal color Doppler in the blood flow studies of postmenopausal women. The *7th Congress of the European Federation of Societies for Ultrasound in Medicine and Biology*, Jerusalem, May 6–11, Abstr. p. 91
46. Kurjak, A. and Žalud, I. (1990). Transvaginal color Doppler. In Kurjak, A. (ed.) *Handbook of Ultrasound in Obstetrics and Gynecology*, Vol.2, p. 305. (Boca Raton, Florida: CRC Press)
47. Kurjak, A., Žalud, I. and Crvenković, G. (1990). The assessment of pelvic circulation by transvaginal color Doppler. *Jpn. J. Med. Ultrasound*, **17**, 116
48. Žalud, I. and Kurjak, A. (1990). The assessment of luteal blood flow in pregnant and non-pregnant women by transvaginal color Doppler. *J. Perinat. Med.*, **18**, 215

12 Uterine Masses

A. Kurjak and I. Žalud

The ability of sonography to depict subtle changes in the myometrium and endometrium makes it the diagnostic modality of choice for the evaluation of many uterine disorders (Figures 12.1–12.5). With sonography, the uterus can be imaged in several scan planes. Once a uterine lesion is suspected clinically, sonography can be used to establish the presence, size, extent and internal consistency of the lesion, as well as to detect associated pathology such as liver metastases. Ultrasound has a major role in differentiating palpable uterine masses from those that arise from adnexal structures (Figure 12.6). The specific diagnosis can be confirmed by endometrial biopsy, thorough dilatation and curettage, by other imaging techniques such as hysterosalpingography, and, in some cases, even by direct hysteroscopic visualization. However, all of these diagnostic procedures are too invasive to be applied in routine clinical practice. This chapter discusses and illustrates the possibilities of transvaginal color Doppler sonography in non-invasive tissue characterization of uterine masses.

BENIGN UTERINE MASSES

Endometriosis, endometrial or Nabothian cysts disturb normal uterine architecture (Figures 12.7 and 12.8). However, myoma (fibroma) is the most common benign uterine mass which consists of smooth muscle and connective tissue. Actually, it is estimated that myomas are present in 20% of women over 35 years of age[1]. They are estrogen-dependent and usually regress after menopause. Myomas are most prevalent in black women and other dark-skinned groups. The usual clinical finding is a palpable mass in a middle-aged woman, but myoma may be associated with excessive menstrual bleedings and pelvic pain.

Myomas or fibroids are important to the sonologist for two reasons. First, as a neoplasm, a myoma does have a small malignant potential. More commonly, however, the clinician needs to differentiate a palpable mass of uterine origin from an adnexal mass. For these reasons, detailed sonographic examination of these tumors is warranted. Myomas usually develop in the myometrium of the upper contractile fundal and corporeal portions of the uterus. Only 3% are of cervical origin. Intramural myomas cause the uterus to contract, and the resultant compression of these tumors is believed to displace them either toward the peritoneal surface to form subserous nodules or toward the endometrial cavity to produce submucous nodules. Myomas are usually multiple and of various sizes. A solitary nodule is found in only 2% of patients[1,2]. As the tumor increases in size, it may eventually outgrow the blood supply and central ischemia is followed by various stages of degeneration.

Transabdominal ultrasound

The typical sonographic appearance of myoma consists of a mildly to moderately echogenic intrauterine mass that causes nodular distortion of the uterine outline. Small intramural or submucous myomas may be recognized by their distortion of the normally linear central endometrial echoes (Figure 12.9). The solid nature of a fibroid may often cause an indentation of the bladder or rectum (Figure 12.10). The echogenicity of a fibroid depends upon the relative ratio of fibrous tissue to smooth muscle. With a more fibrous component, there is increased echogenicity of the nodule (Figures 12.11 and 12.12). The sonographic texture of fibroids also depends on the type and presence of degeneration and upon the vascular supply (Figure 12.13). The most common cause of calcification within the uterus is calcific degeneration within a fibroid (Figures 12.14 and 12.15). Of one series of 75 cases of fibroids, 25% had calcifications[1]. The pattern of calcification varied from a few small foci to a large rim of globular calcification. If the calcification is extensive and is located along the anterior portion of the fibroid, it may prohibit complete sonographic delineation of the mass. Other types of degeneration within myomas, that produce sonographically recognizable changes in uterine texture, include cystic, myxomatous, and hyaline degeneration. Myomas that are pedunculated can be confused with

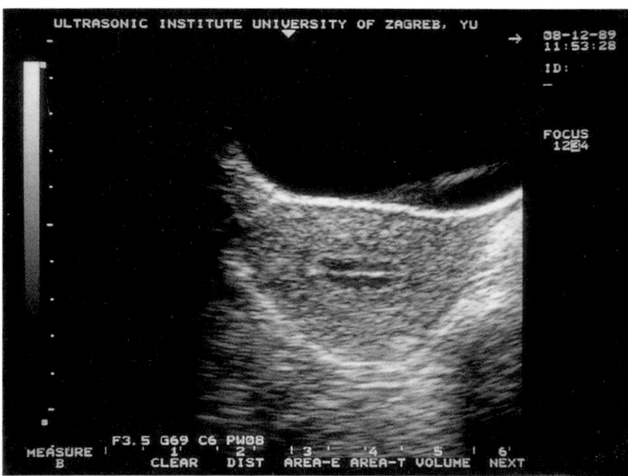

Figure 12.1 A transverse transabdominal sonogram of the normal uterus. The contour is clearly outlined and the texture displays homogeneous and moderate echogenicity. The central cavity echo is clearly visible

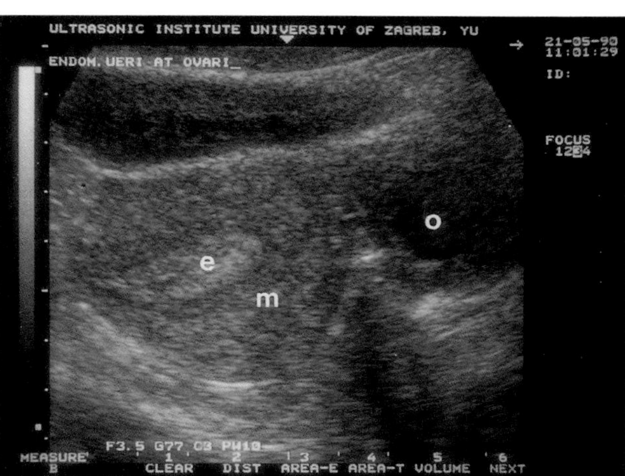

Figure 12.2 An oblique scan of the uterus showing differing echogenicity of the myometrium (m) and endometrium (e). The left ovary (o) can also be visualized

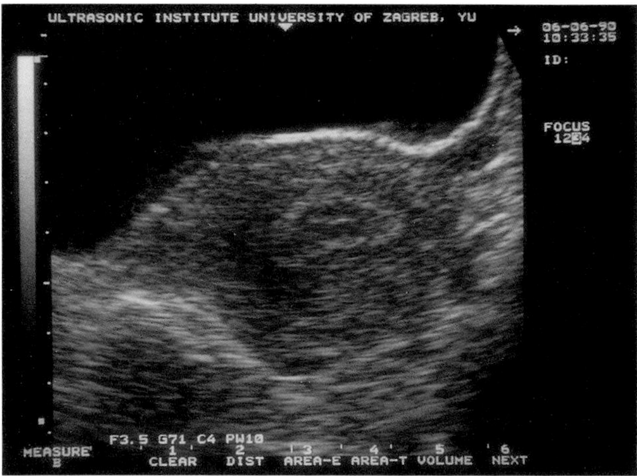

Figure 12.3 Another example of the normal uterine sonogram

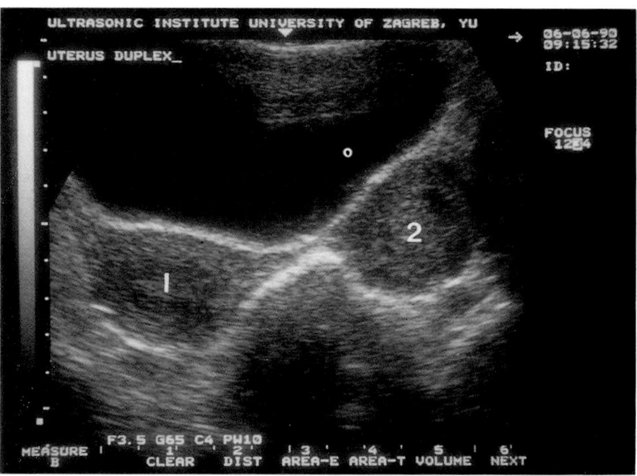

Figure 12.4 Transabdominal scan of a uterus duplex. Two separate uterine bodies and two uterine cavities are demonstrated. 1 = right uterus; 2 = left uterus

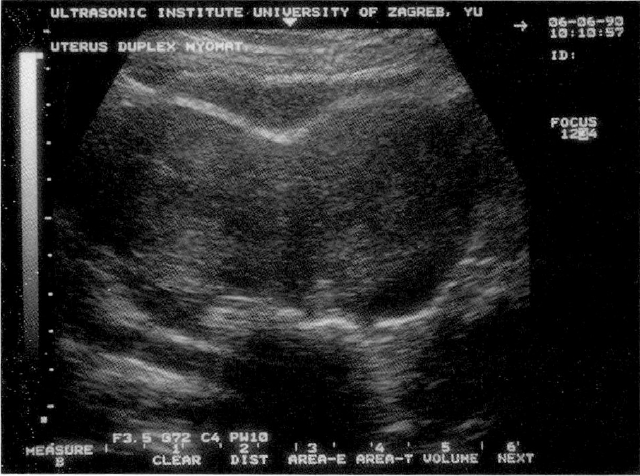

Figure 12.5 A uterus duplex. An irregular contour and unhomogenicity of the uterine texture can be seen in the case of a uterine myoma

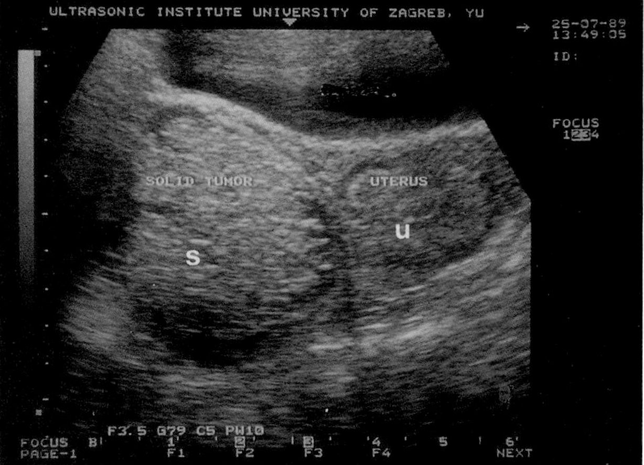

Figure 12.6 A large pelvic mass with secondary changes: degeneration, calcification and necrosis. The texture of the tumor is unhomogeneous and exhibits a mixture of strong echoes and irregular hypoechoic areas. s = solid tumor; u = uterus

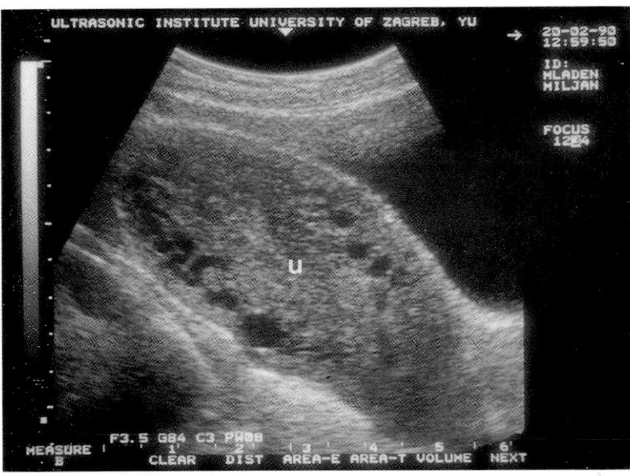

Figure 12.7 Adenomyosis. The homogenicity of the uterine texture is affected by small cystic areas which contain blood. u = uterus

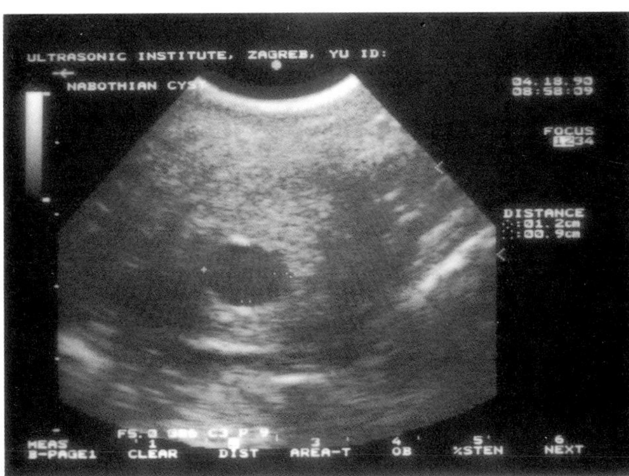

Figure 12.8 A Nabothian cyst. The small cystic structure was detected in the cervical part of the uterus

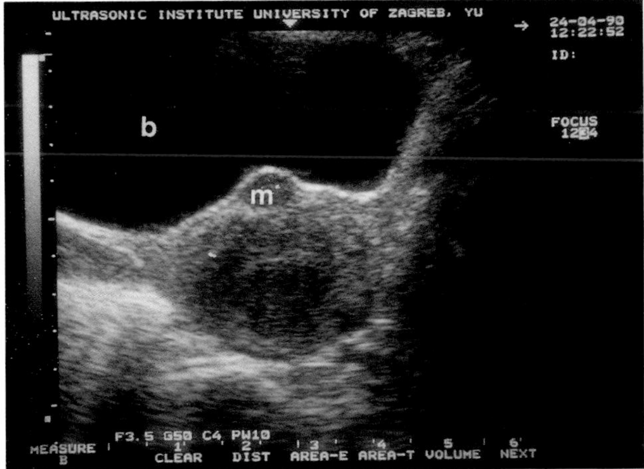

Figure 12.9 A small subserous myoma. Distortion of the normal uterine contour is visible. m = myoma; b = bladder

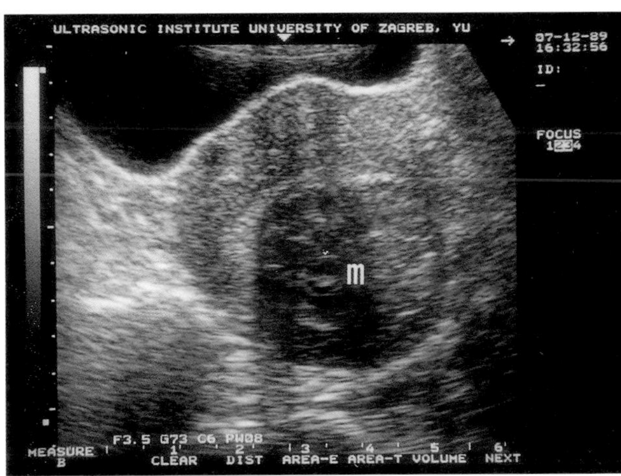

Figure 12.10 A necrotic myoma in the posterior uterine wall. m = myoma

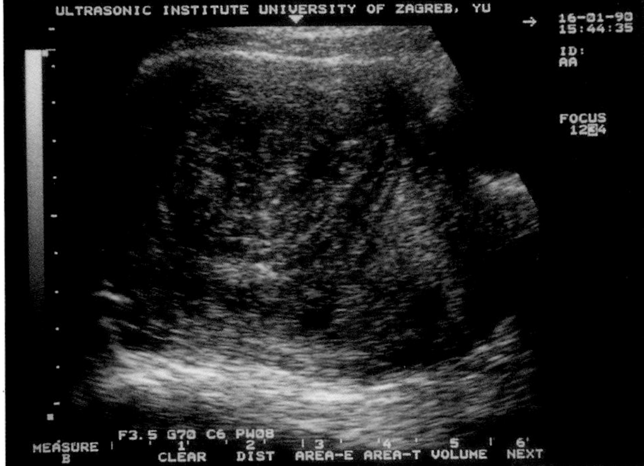

Figure 12.11 Another example of an enlarged myomatosus uterus

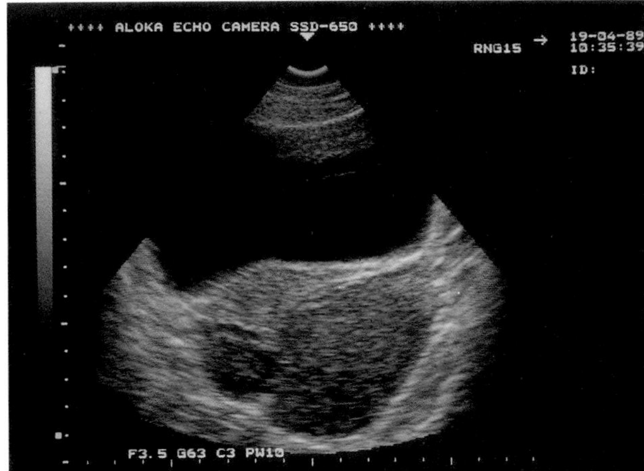

Figure 12.12 An intramural myoma in the right uterine wall. The tumor appears as a cyst because of tumor tissue degeneration and necrosis

other adnexal masses if their pedicle is not visualized. The most common location of pedunculated myomas is superior to the uterine fundus. Furthermore, solid masses that are adjacent to the uterus occasionally appear as masses within the uterine contour[3,4]. In this setting, a retrouterine mass may be misdiagnosed as an enlarged uterus. The most common solid masses to simulate the sonographic appearance of a fibroid are the solid ovarian tumors.

Adenomyosis of the uterus is a condition in which clusters of endometrial tissue occur deep within the myometrium. During menstruation, significant pain can be caused by bleeding of the endometrial tissue within the muscle of the uterus. The uterus may be slightly enlarged. Occasionally, this condition can be diagnosed sonographically because of a thickened and 'swiss cheese' appearance of the myometrium due to areas of hemorrhage and clot within the muscle (Figure 12.16).

Endometrial hyperplasia typically follows prolonged endogenous or exogenous estrogenic stimulation. Adenomatous hyperplasia is thought to be a precursor of endometrial carcinoma[5]. Endometrial hyperplasia may be suspected sonographically because of an abnormal thickening (greater than 6 mm) of the endometrium. Pseudopolyps of endometrial tissue may appear to project into the uterine cavity.

In cases of uterine masses, the use of abdominal sonography enables one to investigate only the suspected area, if anamnesis and results of clinical investigations are available. The borders of tumor spread can only be detected if the tumor extends beyond the boundary of the organs.

Transvaginal sonography

Like other non-invasive methods, transvaginal sonography does not differentiate between benign and malignant lesions accurately. However, in cases of known histological diagnosis of cancer, localization and tumor extent can be imaged more precisely[6]. The numerous advantages of the transvaginal approach have led to an improved pretherapeutical staging[7-10]. A better survey of the true pelvis is guaranteed and the intracorporeal position of the scanner gives the possibility of using higher frequencies which improve resolution capacity. The scanner is positioned close to the uterus. For each patient, the small distance between the transducer and the target organs is nearly equal, and the investigation can be performed under standardized settings of the ultrasound equipment. Additionally, embarrassing layers located in the beam between the transabdominal probe and the target organ, such as filled loops of intestine and bone structures, can be excluded by transvaginal sonography. Therefore, any 'blind area' can be excluded. This is particularly important in patients with severe adhesions after surgery or radiation therapy, or in cases of inflammation or endometriosis. Transabdominal scanning may be hindered by obesity or in the case of an empty bladder, whereas transvaginal sonography is not affected by these circumstances.

Doppler ultrasound

Conventional B-mode transvaginal sonography has obviously improved accuracy of ultrasound diagnosis of uterine masses. However, the nature of a uterine mass is one of the most important parameters in the assessment of the gynecological patient. The question, 'Is it a benign or a malignant uterine tumor?', cannot be answered by using only B-mode sonography. Morphological features may help clinicians to support the presumptive diagnosis, but controversy still remains. Some benign uterine masses have been proved to be malignant and vice versa. Morphological criteria alone are not good enough in the characterization of uterine masses[11-17].

Neovascularization is well known to the pathologist but recently such a term has also been used by the sonographist (see Chapter 10). It was impossible to see the small vessels which feed growing tumor tissue or analyze blood flow until transvaginal Doppler came into use[11-18]. The presence of uterine tumor vascularity might be used in tissue characterization, similar to the characterization of adnexal masses. Hypervascularity can easily be detected by transvaginal color Doppler (Figure 12.17). Abundant color flow serves as a guide for pulsed Doppler study. Waveform analysis has been used to quantify the resistance of blood flow in tumor vessels (Figure 12.18). The results obtained, as was extensively discussed in the case of adnexal masses, are extremely important in the differentiation between normal existing and newly formed vessels. This is a key for clinical '*in vivo*' non-invasive assessment of the nature of tumors. The Zagreb group has been a pioneer in such an approach[19]. They noticed that the vascularization of benign uterine masses is supported by normal existing vessels. When color flow was present, waveform analysis indicated blood flow similar to the normal myometrial perfusion originating from the terminal branches of the uterine artery (Figure 12.19). The diastolic flow was always present, usually increased in comparison with the uterine artery blood flow, but the Pourcelot resistance index (RI) was moderately high to very high ($RI > 0.50$) (Table 12.1). Such a vascular signature was different from cases of uterine malignancy and therefore was used in clinical assessment of benign

Table 12.1 Transvaginal color Doppler assessment of benign uterine masses (n = 201)

Histopathology	*n*	*Color flow* *n*	*Color flow* *%*	*RI (2 SD)*
Myoma	156	61	39.1	0.59 (0.08)
Adenomyosis	42	17	40.5	0.67 (0.10)
Endometriotic cyst	3	0	—	—

uterine masses. However, the ability to demonstrate tumor vessels within uterine fibromas is largely dependent on the tumor size, and the position and extent of secondary degenerative changes (Figures 12.20–12.26). The vessels observed in benign tumor tissue were typically thin with low-velocity flow. As further clinical trials show that the analysis of blood flow supply can be used as a sufficiently reliable criterion for evaluation of the nature of uterine masses (benign vs. malignant), the impact of transvaginal color Doppler diagnosis in the management of gynecological patients with uterine tumors will truly be a major one. It is also postulated that such vascular signatures might be of great help in predicting the growth rate of benign uterine masses.

MALIGNANT UTERINE MASSES

Carcinoma of the uterine body is a very different disease from carcinoma of the cervix. Adenocarcinoma of the uterus is typically a disease of perimenopausal women[20]. Epidemiologically, these patients tend to have had fewer pregnancies and a later first pregnancy than patients with cervical lesions. An association with estrogen therapy has been strongly suggested. Occasionally, endometrial cancer can occur in young women. Clinically, adenocarcinoma of the endometrium typically presents with a watery or bloody discharge. These tumors of the uterine corpus spread either by local invasion to surrounding organs, such as bladder and rectum, or by lymphatic spread.

Transabdominal ultrasound

In B-mode ultrasonic terms, uterine malignancy can hardly be distinguished from fibromas. Typical findings include uterine enlargement and non-homogeneous tumor texture due to necrosis and other secondary changes (Figures 12.27–12.30)[21–23]. When carcinoma first occurs, a decidual reaction, a line of echoes with a sonolucent center in the endometrial cavity, may be seen. Fluid within the endometrial cavity is another early sign (Figure 12.31). When more advanced, the uterus takes on an irregular border without the rather smooth lobulation usually seen with fibroids; areas of neoplastic involvement are hypoechoic. Since fibroids are so common, it is not unusual for fibroids and carcinoma of the body of the uterus to coexist. Carcinoma, as opposed to a fibroid, is almost always hypoechoic. There is a bulge away from the main body of the uterus and spread is eccentric.

Carcinosarcomas are uncommon tumors of mixed origin, containing stromal and Müllerian elements. Patients tend to be in their fifties and sixties and are postmenopausal. The uterus is usually enlarged two to three times normal and is filled with polypoid masses. Sarcoma botryoides is a rare tumor arising from the uterus in children. It is a mixed stromal tumor, predominantly rhabdomyosarcoma. Sarcoma botryoides is a soft, polypoid mass, frequently large, and often extruding from the cervix. Ultrasonographically, it appears as a complex mass deforming the uterine outline[24]. Leiomyosarcoma can arise from a pre-existing leiomyoma or from muscle or connective tissue within myometrium or blood vessels (Figures 12.32 and 12.33). The tumor may be too small to be seen or may be indistinguishable from a benign leiomyoma. Ultrasound findings are non-specific.

Transvaginal sonography

Transvaginal sonography provides a better visualization of the myometrium and endometrial echoes[9,25]. It can be performed immediately after bimanual palpation or, if necessary, simultaneously. The hand on the abdominal wall may also guide the uterus closer to the scanner. The vaginal approach allows the measurement of the extension of the uterus length and of the endometrial echoes. Differentiation and localization of myometrial infiltration are also possible. Pathological proliferation of the endometrium may be visualized by the vaginal transducer a long time before symptoms are apparent. In cases of endometrial cancer, the echoes of the endometrium are thickened and inhomogeneous, and no liquid is visible in this area as in the case of mucometria. Advanced carcinomas of the endometrium always show a pathologically thickened endometrium with typical 'fuzzy', inhomogeneous margins toward the myometrium and a typical echo-poor area of carcinoma invasion into the myometrium. The boundary towards the myometrium is not sharp. In contrast, myomas show a clearly recognizable delineation and are separated by a capsule. Since the corpus uteri and also the cervix can be imaged at the same time, an additional invasion of the cervical area can be ascertained. The depth of an endometrial carcinoma, its longitudinal dimension, and the

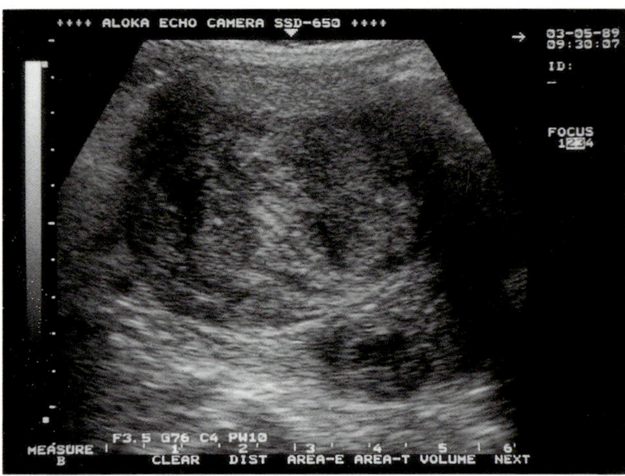

Figure 12.13 A uterine myoma. The sonographic appearance of the tumor depends on the type and presence of degeneration

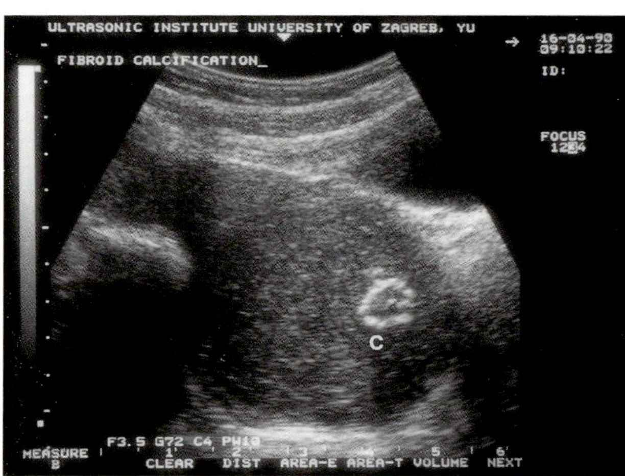

Figure 12.14 The typical ultrasound finding of fibroid calcification (c)

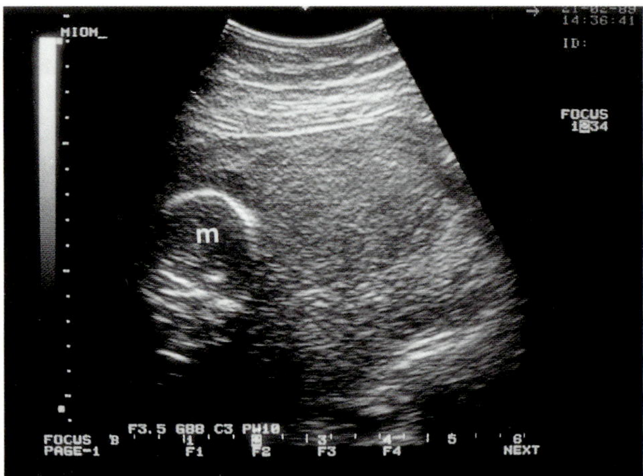

Figure 12.15 Another example of tumor calcification. The border of the myoma (m) is clearly outlined

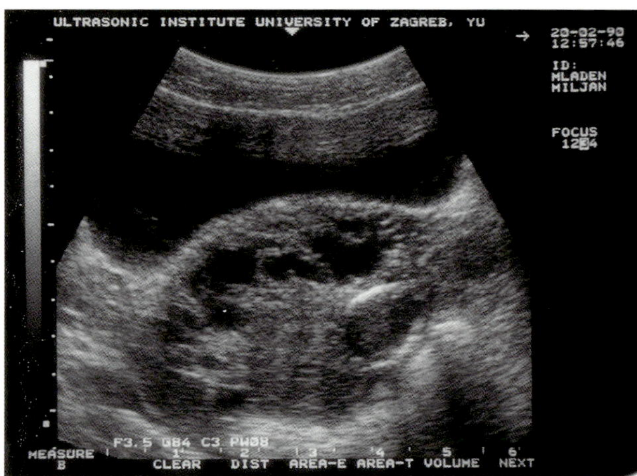

Figure 12.16 Adenomyosis uteri. A large adenomyoma can be observed in the anterior and lateral uterine walls. The uterus is markedly hypoechoic and several large cysts are visible

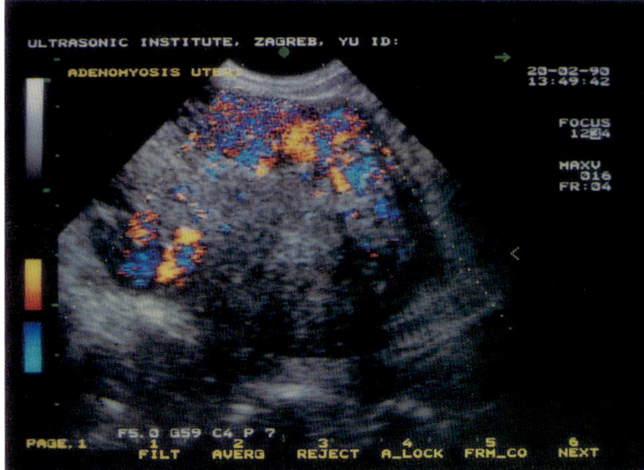

Figure 12.17 Adenomyosis uteri. Transvaginal color Doppler showing prominent vascularization of the uterine tissue

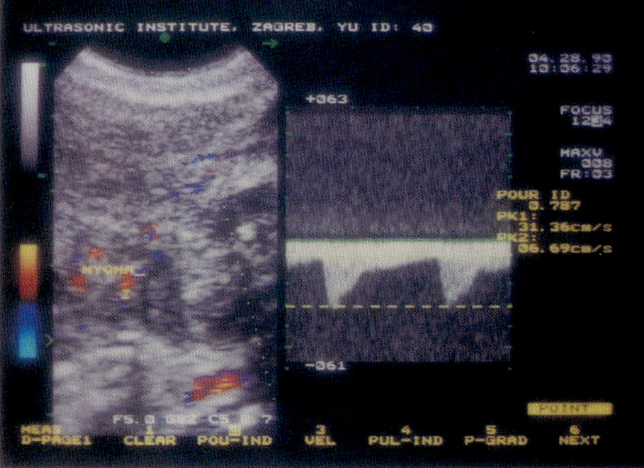

Figure 12.18 Color Doppler indicating blood flow at the periphery of a uterine myoma. Color signals were explored by pulsed Doppler. Waveform analysis (right) shows moderate-velocity and high-resistance blood flow. Benign nature of the tumor was confirmed by histopathology

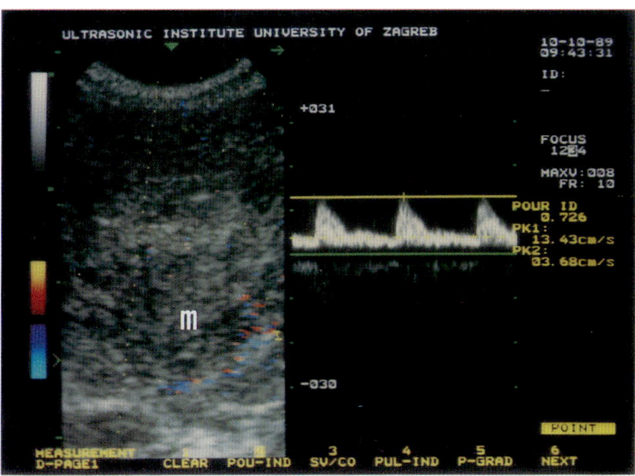

Figure 12.19 A uterine myoma (m). Color Doppler detected increased uterine perfusion. Pulsed Doppler (right) shows blood flow similar to the normal myometrial perfusion originating from the terminal branches of the uterine artery

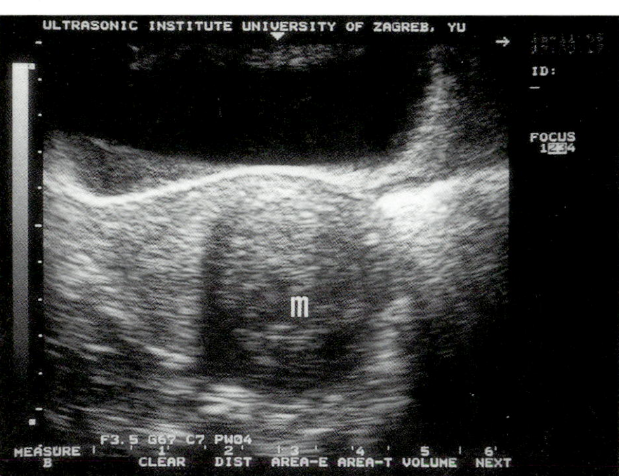

Figure 12.20 The transabdominal scan showing secondary changes in a large uterine mass suspected to be a myoma (m)

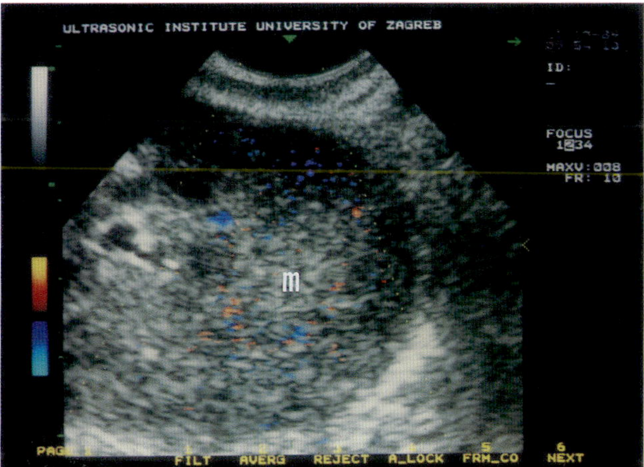

Figure 12.21 The same patient. Transvaginal color Doppler detected a good vascular supply to the uterine tumor. m = myoma

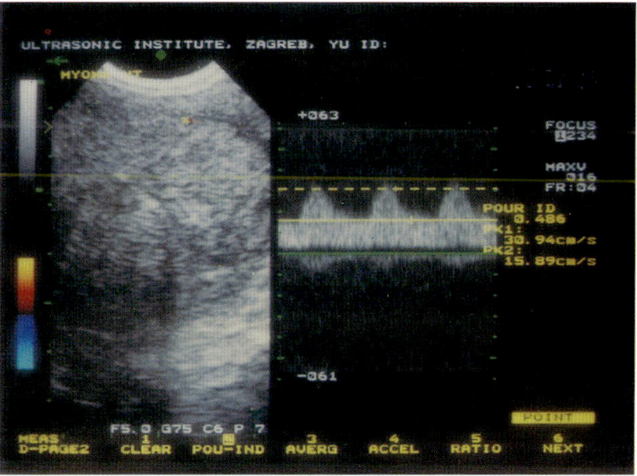

Figure 12.22 The same patient. The waveform analysis (right) indicates moderate-velocity and low-resistance blood flow. The diastolic component of cardiac cycle was increased. A borderline Doppler finding. The benign nature of the uterine mass was confirmed by histopathology

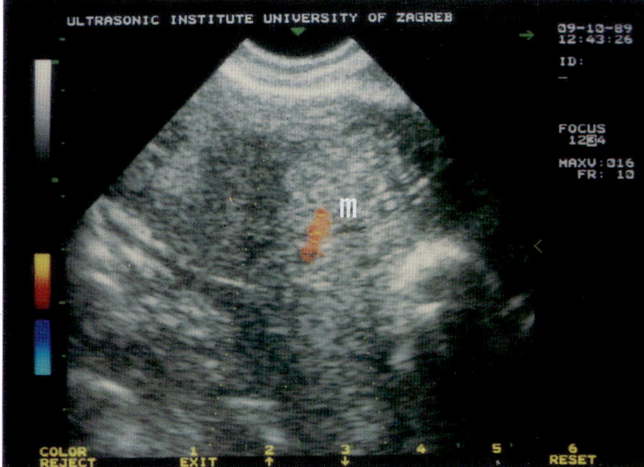

Figure 12.23 Solitary tumor vessel detected by color Doppler. Prominent color signal indicated a high velocity of blood flow. m = myoma

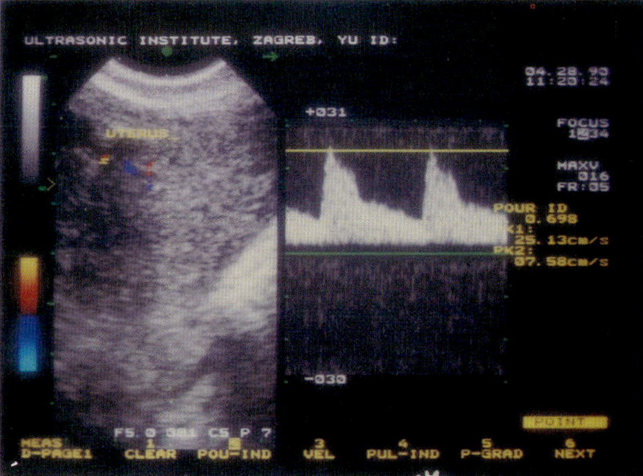

Figure 12.24 Color flow obtained on the periphery of the solid uterine mass suspected to be a myoma. Pulsed Doppler (right) showed moderate-velocity and moderate-resistance (RI = 0.698) blood flow. A myoma was diagnosed by histopathology

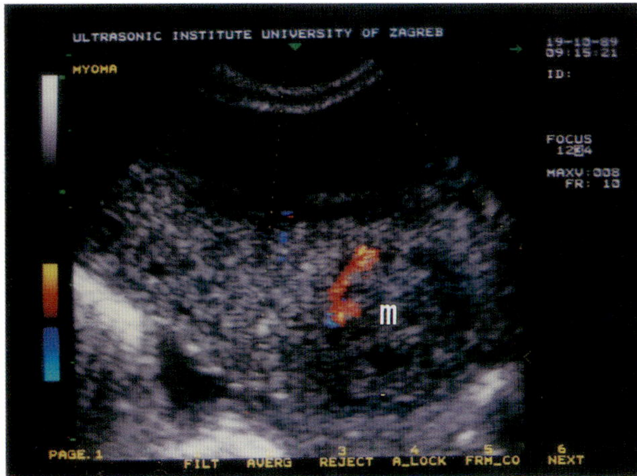

Figure 12.25 Another example of solitary tumor vessels visualized by color Doppler. m = myoma

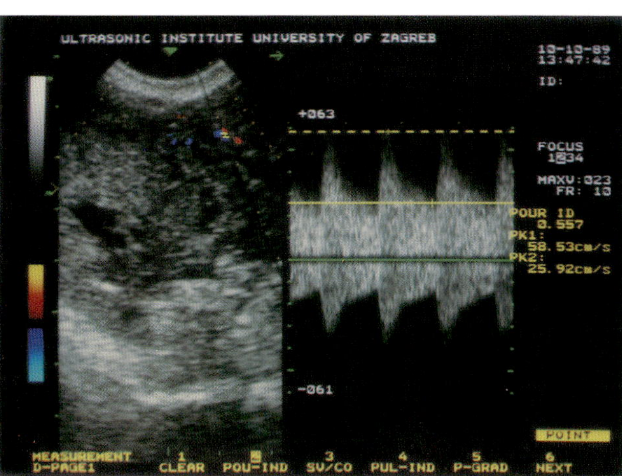

Figure 12.26 Color and pulsed Doppler indicating a benign uterine mass

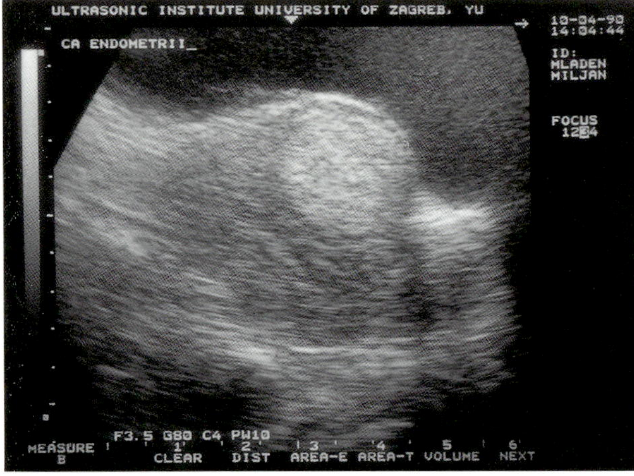

Figure 12.27 A transabdominal scan of extremely unhomogeneous uterine tissue. Endometrial cancer was suspected

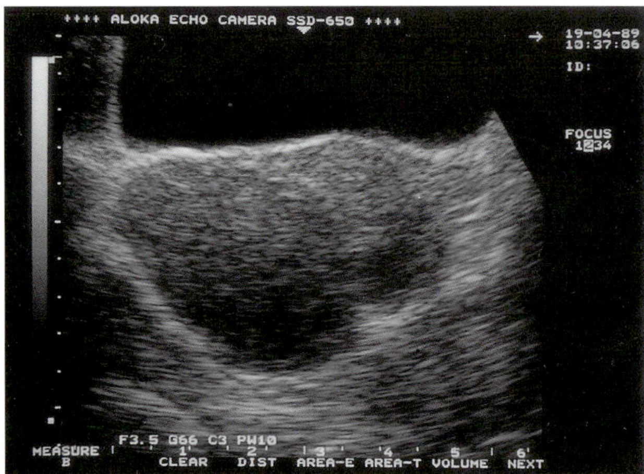

Figure 12.28 A large solid uterine mass which occupied almost all the uterine tissue

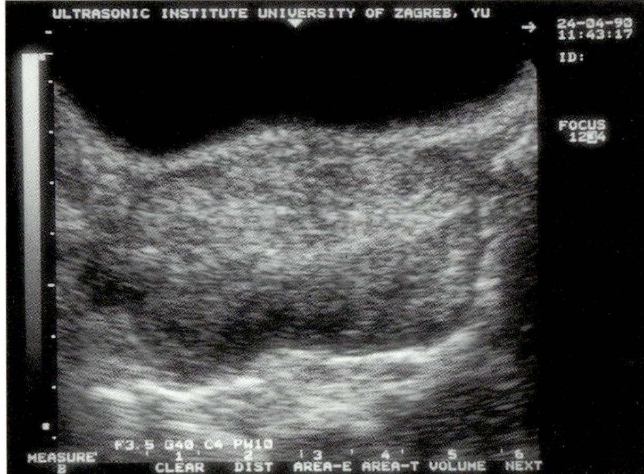

Figure 12.29 A transabdominal scan of an enlarged uterus. The texture of the tumor is irregular and almost half of the tumor tissue shows signs of necrosis and degeneration

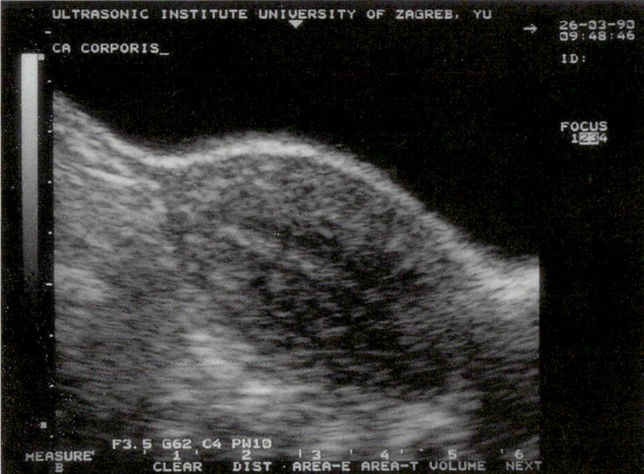

Figure 12.30 The enlarged and unhomogeneous uterus was visualized by transabdominal ultrasound

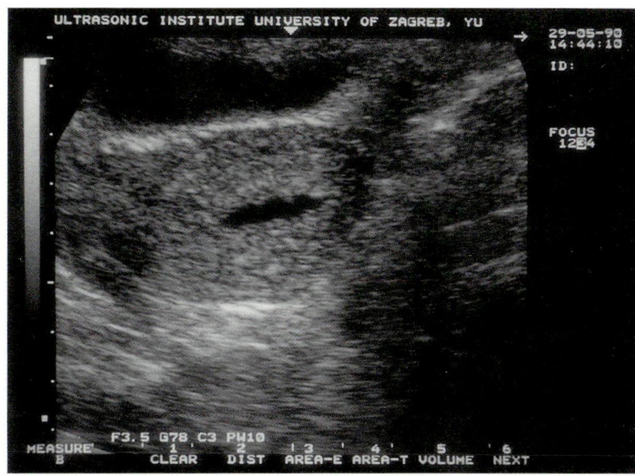

Figure 12.31 A line of echoes with a sonolucent center can be seen in the endometrial cavity as an early sign of endometrial cancer

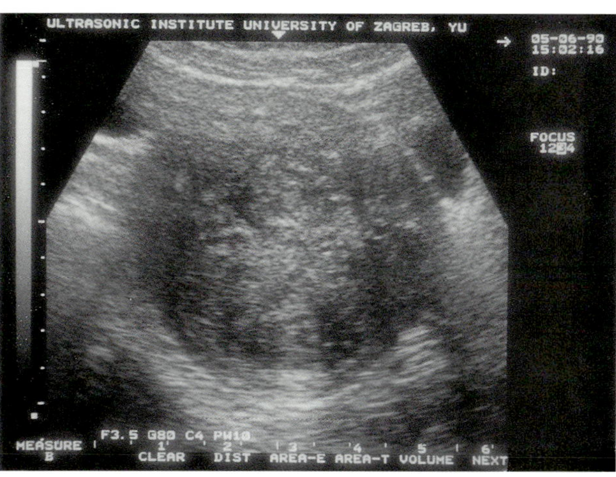

Figure 12.32 An extremely large uterine tumor with pronounced degeneration, necrosis and hemorrhage. The diagnosis of uterine sarcoma was confirmed on laparotomy

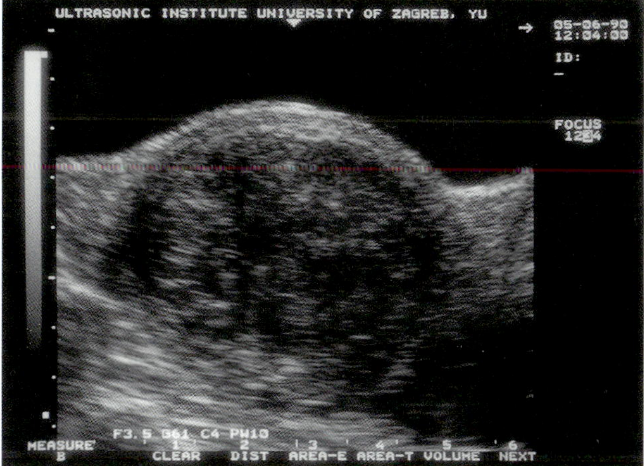

Figure 12.33 Another example of ultrasound finding suspected to be a uterine sarcoma

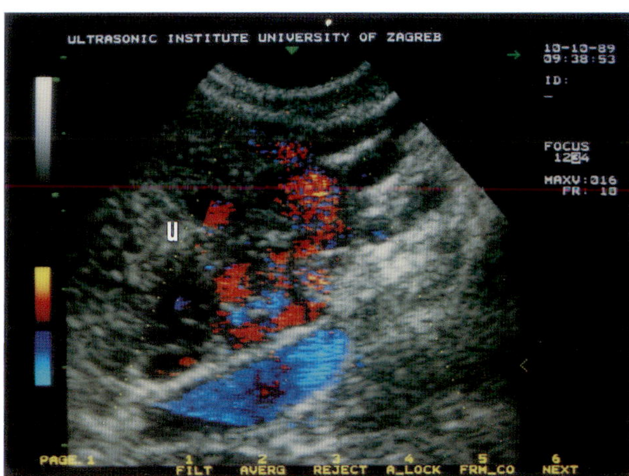

Figure 12.34 Transvaginal color Doppler showing highly vascularized tumor tissue. u = uterus

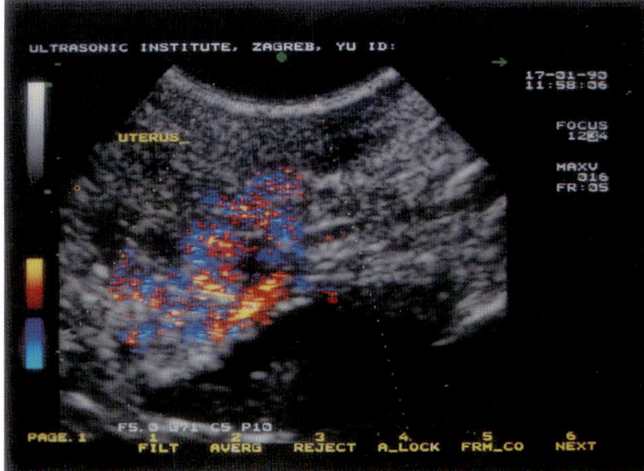

Figure 12.35 Uterine adenocarcinoma. The typical finding was the presence of irregular, thin, randomly dispersed vessels visualized by color Doppler

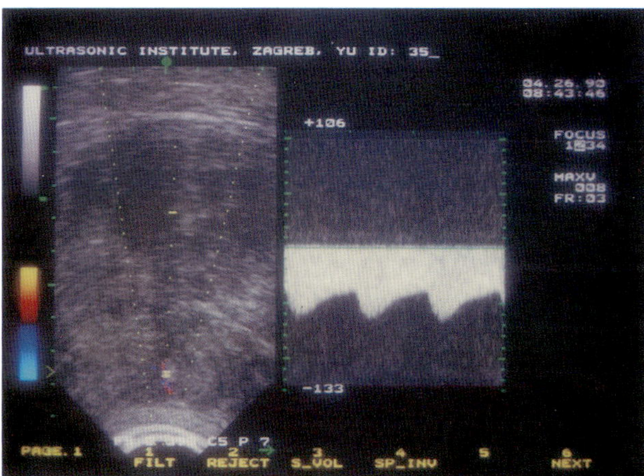

Figure 12.36 Small vessels detected by color Doppler on the periphery of the uterine mass. Pulsed Doppler (right) shows very small systolic–diastolic variations of blood flow. This was suspicious of a malignant uterine tumor, which was later confirmed by histopathology

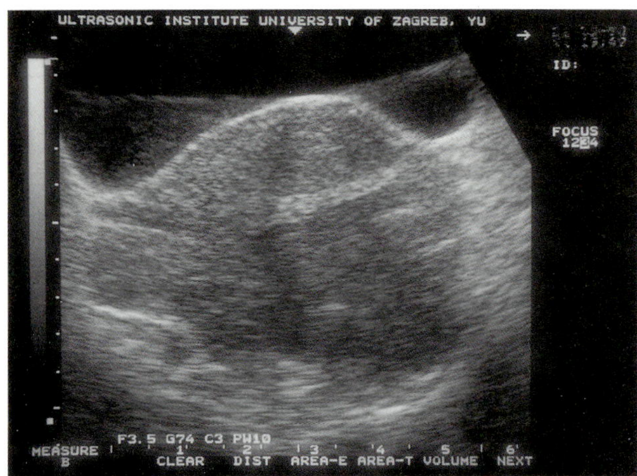

Figure 12.37 A transabdominal scan of the large uterine mass which disturbed normal uterine tissue

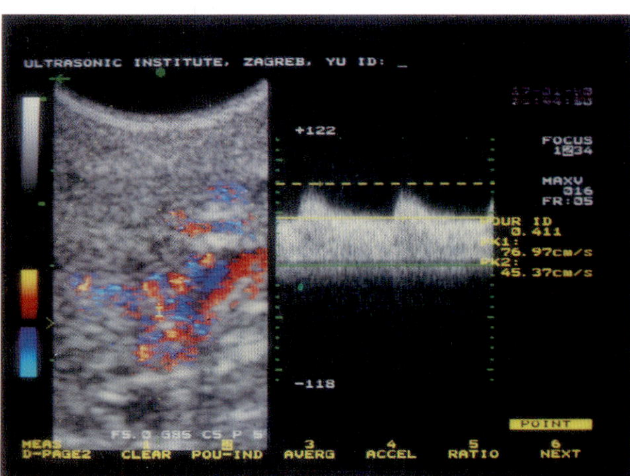

Figure 12.38 The same patient was examined by transvaginal color Doppler. Abundant color signals were obtained from the tumor tissue. Waveform analysis (right) shows high velocity and very low resistance of the tumor blood flow. A typical finding for highly suspicious malignant tumor

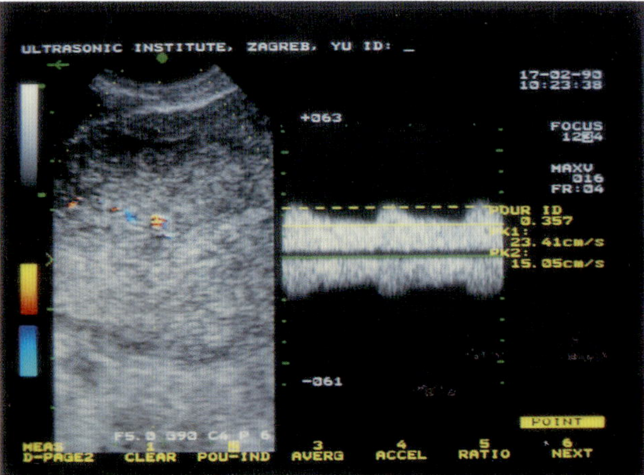

Figure 12.39 A large solid uterine mass. Small tumor vessels are detected by color Doppler. Pulsed Doppler (right) shows low-velocity and very low-resistance (RI = 0.357) blood flow. Endometrial cancer was confirmed by histopathology

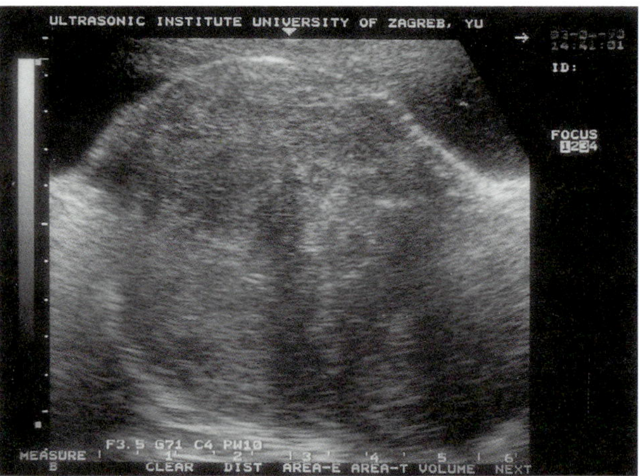

Figure 12.40 The transabdominal scan of a huge uterine tumor. Unhomogenicity was produced by tumor degeneration and hemorrhage

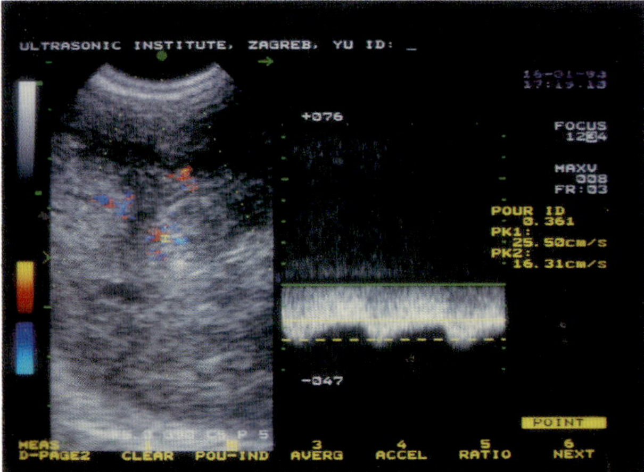

Figure 12.41 The same patient. Transvaginal color and pulsed Doppler indicate the malignant nature of the observed tumor tissue

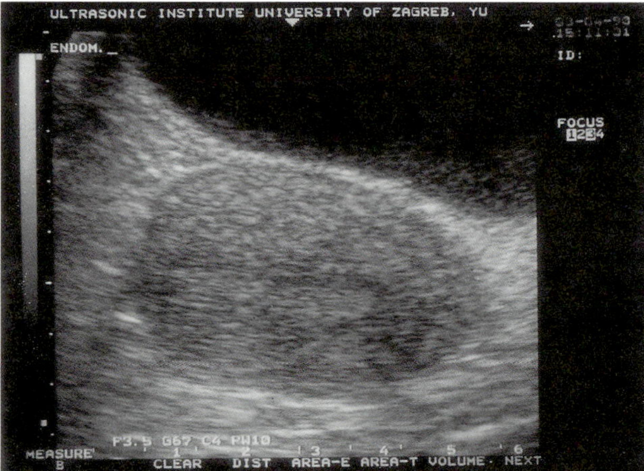

Figure 12.42 Transabdominal ultrasound. A slightly enlarged uterus and an irregular endometrial echo were noted

possible occupation of the cervical region can be evaluated. This allows for preoperative planning of the extent of intervention.

Adenocarcinoma of the endometrium may present with atrophy, normality or enlargement of the uterus and with a normal or abnormal configuration. There is a benign process that can enlarge the uterus coincidental to carcinoma of the endometrium. Local sonographic enlargement due to a tumor may be detectable. A significant percentage of the patients in one series had carcinoma of the endometrium in association with myoma[1]. It is, therefore, difficult to evaluate the extent of the tumor based on the sonographic findings of size or contour alone. Distortion or deviation of the endometrial layer can occur with the development of a neoplasm. But again, this may be due to coincidental fibroids. Endometrial carcinoma generally involves the endometrial surface without significant deviation of this layer. In advanced lesions, the endometrium appears thickened (greater than 6 mm) and markedly irregular. Hypoechoic areas within the lumen may be seen with hematometria. Since early carcinomas are superficial lesions, it is impossible to detect texture changes. Carcinoma of the endometrium can occur anywhere along the length of the endometrial canal, and the area of involvement does not necessarily correlate with the area of ultrasonic change. Areas of hemorrhage surrounding the tumor focus can mimic the area of the actual tumor.

Doppler ultrasound

Obviously, there is some limitation to ultrasound diagnosis and evaluation of malignant uterine masses, particularly in the initial stage of the tumor. Additional diagnostic procedures are required before the diagnosis of uterine malignancy is reached. The Pap smear is absolutely superior to ultrasound in cases of cervical carcinoma. Early diagnosis of endometrial carcinoma still remains a clinical challenge. Even though many morphological features studied by ultrasound have been described and certain clinical symptoms are known, an accurate diagnostic procedure is not defined and universally accepted. Ideas about tumor angiogenesis have recently been explored and carefully studied. Transvaginal color Doppler encourages such attitudes (Figure 12.34). The results obtained in the Ultrasonic Institute in Zagreb[26] have confirmed Folkman's hypothesis that a malignant rapidly growing tumor produces its own vessels[27]. Such vessels contribute to increased blood flow perfusion, which is an excellent background for tumorigenesis. Our group has investigated the vascular supply of different uterine masses in order to assess the criteria for early detection of uterine malignancy and to improve the accuracy of ultrasound in differentiation between benign and malignant uterine masses[11–13, 16, 17, 26]. An abnormal blood flow pattern has been noticed in all cases of endometrial adenocarcinoma and uterine sarcoma (Table 12.2). The typical finding was the presence of irregular, thin, randomly dispersed vessels (Figure 12.35). The major advantage of color Doppler is the spatial display of blood flow, which provides for easy orientation concerning vascularity in certain scanning areas, and facilitates a detailed analysis of flow velocity patterns by pulsed Doppler. This is particularly advantageous in the analysis of flow in thin vessels which are extremely difficult to identify by pulsed Doppler only (Figure 12.36). The amount of color flow has varied very much and corresponded to the number of the vessels and velocity of blood flow through them. Such a signature is very probably correlated with the temporary condition of the tumor and may reflect the stage and type of the malignant tumor (Figures 12.37–12.43). If so, transvaginal color Doppler is going to play the major role in the non-invasive characterization of uterine malignancy and may directly influence the therapeutic procedure. In the case of operative or radiological therapy, color Doppler can help in the effective follow-up of the current process. But the success rate of color Doppler in the diagnosis of uterine malignancy was disturbed by cervical carcinoma. No patient examined for cervical carcinoma *in situ* presented abnormal vascular signs. It is true that this normally highly vascularized area is a good pre-existing condition for tumor growth. It is also speculated that newly formed vessels in this stage of cervical carcinoma are too small and the velocity and volume flow are below the resolution power of the equipment. Hata *et al.* could not detect any pulsed Doppler signal in 50% of advanced cervical cancer cases, and an abnormal finding was present in only 16.7% of these patients[28]. Furthermore, in all cases of endometrial carcinoma and ovarian malignancy, typical abnormal flow patterns were

Table 12.2 Transvaginal color Doppler assessment of malignant uterine masses (n = 31)

		Color flow		
Histopathology	*n*	*n*	%	*RI (2 SD)*
Adenocarcinoma	12	12	100	0.34 (0.05)
Leiomyosarcoma	3	3	100	0.31 (0.03)
Sarcoma botryoides	1	1	100	0.33
Cervical carcinoma *in situ*	15	0	—	—

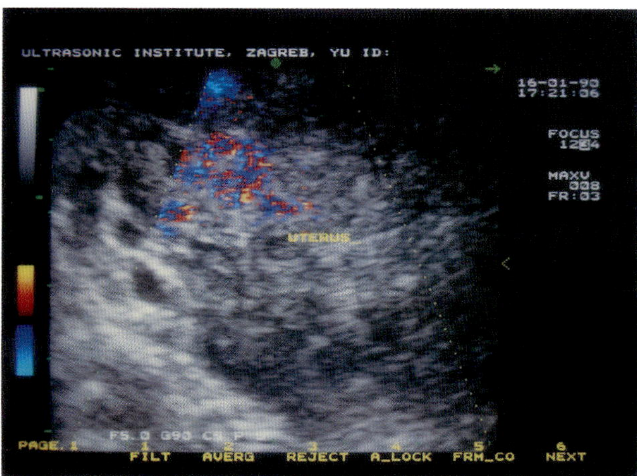

Figure 12.43 The same patient. Small tumor vessels are visualized by transvaginal color Doppler. The malignant nature of the uterine mass was also confirmed by pathology

noted. Moreover, a decrease in blood flows was observed in most subjects after chemotherapy by anticancer drugs or irradiation.

CONCLUSION

Transvaginal color Doppler is a pertinent and non-invasive diagnostic tool that can be used repeatedly for assessing tumor vascularity in uterine disorders. Present experience has shown that the use of only B-mode ultrasound examination is not sufficient to predict the nature of the tumor. Doppler ultrasound obviously increases the reliability of ultrasound diagnosis in certain uterine pathologies (myoma, endometrial carcinoma). The opportunity to visualize and analyze blood flow characteristics in uterine vessels will undoubtedly provide new data on the physiology and pathophysiology of uterine perfusion. The diagnostic tool to accurately provide *in vivo* differentiation between benign and malignant uterine masses has been out of clinical reach for a long time. Color Doppler seems to have great potential for this purpose.

REFERENCES

1. Fleischer, A.C., Entman, S.S., Porrath, S.A. and James, A.E. (1985). Sonographic evaluation of uterine malformations and disorders. In Sanders, R.C. (ed.) *The Principles and Practice of Ultrasonography in Obstetrics and Gynecology*, p. 531. (Norwalk: Appleton Century Crofts)
2. Kurjak, A. (1986). *Atlas of Ultrasonography in Obstetrics and Gynecology*, p. 245. (Zagreb: Mladost)
3. Bowie, J. (1977). Ultrasound of gynecologic pelvic masses: the indefinite uterus sizes and other patterns associated with diagnostic error. *J. Clin. Ultrasound*, **5**, 323
4. Kurjak, A. and Žalud, I. (1990). Female pelvis. In Kurjak, A. and Fuckar, Z. (eds.) *Atlas of Abdominal and Small Parts Ultrasonography*. (Zagreb: Naprijed) in press
5. Johnson, M., Graham, M. and Cooperburg, P. (1982). Abnormal endometrial echoes: sonographic spectrum of endometrial pathology. *J. Clin. Ultrasound*, **1**, 181
6. Bernaschek, G. and Deutinger, J. (1990). Endosonography for the diagnosis of uterine malignomas. In Kurjak, A. (ed.) *Handbook of Ultrasonography in Obstetrics and Gynecology*. (Boca Raton, Florida: CRC Press)
7. Timor-Tritsch, I.E., Rottem, S. and Boldes, R. (1988). Scanning the uterus. In Timor-Tritsch, I.E. and Rottem, S. (eds.) *Transvaginal Sonography*, p. 27. (New York, Amsterdam, London: Elsevier)
8. Lewit, N., Thaler, I. and Rottem, S. (1990). The uterus: a new look with transvaginal sonography. *J. Clin. Ultrasound*, **18**, 331
9. Fleischer, A.C., Gordon, A.N., Entman, S.S. and Kepple, D.M. (1990). Transvaginal scanning of the endometrium. *J. Clin. Ultrasound*, **18**, 337
10. Timor-Tritsch, I.E., Rottem, S. and Thaler, I. (1988). Review of transvaginal sonography: a description with clinical applications. *Ultrasound Q.*, **6**, 1
11. Kurjak, A., Žalud, I., Jurković, D., Alfirevic, Z. and Miljan, M. (1989). Transvaginal color Doppler for the assessment of pelvic circulation. *Acta Obstet. Gynecol. Scand.*, **68**, 131
12. Kurjak, A., Žalud, I., Alfirevic, Z. and Jurković, D. (1990). The assessment of abnormal pelvic blood flow by transvaginal color Doppler. *Ultrasound Med. Biol.*, in press
13. Kurjak, A., Jurković, D., Alfirevic, Z. and Žalud, I. (1990). Transvaginal color Doppler imaging. *J. Clin. Ultrasound*, **18**, 227
14. Kurjak, A. (1989). Transvaginal color Doppler in the assessment of pelvic circulation. *Jpn. J. Med. Ultrasound*, **16** (Suppl. II), 1
15. Kurjak, A. and Jurković, D. (1989). Transvaginal color Doppler in the assessment of pelvic masses. Proceedings of *6th World Congress of In Vitro Fertilization and Alternate Assisted Reproduction*, Jerusalem, Abstr. p.26
16. Kurjak, A. and Žalud, I. (1990). Transvaginal color Doppler in the characterization of pelvic masses. *Ultraschall Med.*, in press
17. Kurjak, A. and Žalud, I. (1990). Transvaginal color Doppler. In Chervanak, F.A., Isaacson, G. and Campbell, S. (eds.) *Textbook of Ultrasound in Obstetrics and Gynecology*. (Boston: Little, Brown & Co.) in press
18. Hata, T., Hata, K., Senoh, D., Makihara, K., Aoki, S., Takamiya, O., Kitao, M. and Umaki, K. (1989). Transvaginal Doppler color flow mapping. *Gynecol. Obstet. Invest.*, **27**, 217
19. Kurjak, A. and Žalud, I. (1990). Transvaginal colour flow imaging and ovarian cancer. *Br. Med. J.*, **300**, 330
20. Carson, S.C. (1978). Ultrasound in obstetrics and gynecology. *J. Reprod. Med.*, **20**, 1
21. Sanders, R.C. (1984). Ultrasound in pelvic malignancy. In Steel, W.B. and Cochrane, W.J. (eds.) *Gynecologic Ultrasound*, p. 139. (New York, Edinburgh, London, Melbourne: Churchill Livingstone)

22. Cacciatore, B., Lehtovitra, P. and Wahlstrom, T. (1989). Preoperative sonographic evaluation of endometrial cancer. *Am. J. Obstet. Gynecol.*, **160**, 133
23. Fleischer, A.C., Dudley, B.S., Entman, S.S., Baxter, J.W., Kalemeris, G.C. and James, A.E. (1987). Myometrial invasion by endometrial carcinoma: sonographic assessment. *Radiology*, **162**, 303
24. Woodring, J., Halberg, D. and Daff, D. (1982). Sarcoma botryoides of the uterus presenting as an abdominal mass: a case report. *J. Clin. Ultrasound*, **10**, 347
25. Andolf, E. and Jorgensen, C. (1990). A prospective comparison of transabdominal and transvaginal ultrasound with surgical findings in gynecologic disease. *J. Ultrasound Med.*, **9**, 71
26. Kurjak, A., Žalud, I. and Crvenković, G. (1990). The assessment of pelvic circulation by transvaginal color Doppler. *Jpn. J. Med. Ultrasound*, **17**, 116
27. Folkman, J. (1972). Anti-angiogenesis: new concept for therapy of solid tumors. *Ann. Surg.*, **175**, 183
28. Hata, T., Hata, K., Senoh, D., Makihara, K., Aoki, S., Takamiya, O. and Kitao, M. (1989). Doppler ultrasound assessment of tumor vascularity in gynecologic disorders. *J. Ultrasound Med.*, **8**, 309

13 Ultrasonic Exposures in Transabdominal and Transvaginal Sonography

G. Kossoff

Although it is generally acknowledged that transabdominal gray-scale imaging is unlikely to cause any significant biological effect, concerns with safety continue to be expressed regarding transvaginal gray-scale imaging and color and duplex Doppler examinations. The resolution of the question of safety is a complex task and will take some time to achieve[1]. However, an appreciation of some of the issues involved can be elucidated by analysis of the ultrasonic exposures used in the various techniques. Fortunately, manufacturers today supply information on the acoustic output of their equipment, which makes this analysis a relatively simple task.

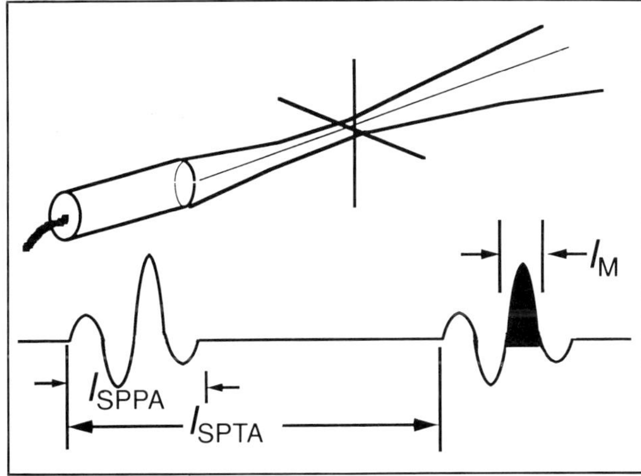

Figure 13.1 Intensity parameters used to specify acoustic output of ultrasonic diagnostic equipment

CURRENT POLICY ON ACOUSTIC OUTPUT

In order to restrict the indiscriminate increase in the acoustic output of new equipment, the Center for Devices and Radiological Health of the United States Food and Drug Administration (FDA) has produced a document entitled *The 510 (k) Guide for Measuring and Reporting Acoustic Output of Diagnostic Medical Devices.* This document quotes the maximum values of three intensity parameters of acoustic output which were generated by equipment manufactured in 1976. It also stipulates a policy decision which states that new gray-scale and/or Doppler equipment may be introduced on the market in the United States without having to undergo investigational approval if the acoustic output of the new equipment is below these maximum values. The effect of this document has been to persuade the majority of the manufacturers to limit the acoustic output of their equipment to comply with the FDA specifications.

The values of the three intensity parameters as measured in water are given in Table 13.1 and they are illustrated schematically in Figure 13.1. The spatial peak-temporal average intensity I_{SPTA} specifies the intensity on the axis averaged over time and thus takes into account the different pulse repetition rates used by the equipment. It is related, through the beam width, to the power content in the ultrasonic beam and the potential thermal elevation which could result during the exposure. The spatial peak-pulse average I_{SPPA} and maximum intensity I_M are the intensities associated with the pulse waveform. I_{SPPA} specifies the axial intensity averaged over the duration of the pulse while I_M specifies the axial intensity averaged over the largest half-cycle in the pulse. Both are relevant to the potential cavitational activity which could occur during the exposure.

Table 13.1 Maximum acoustic output measured in water as specified in the FDA 510 (k) document

Intensity	*Cardiac*	*Vascular*	*Eye*	*Fetal*
I_{SPTA} (mW/cm^2)	730	1500	70	180
I_{SPPA} (W/cm^2)	350	350	110	350
I_M (W/cm^2)	550	550	200	550

Currently, major interest is being expressed in the thermal mechanisms for biological effects of diagnostic ultrasound. The spatial peak-temporal average I_{SPTA} intensity is the relevant parameter. As shown in Table 13.1, the maximum allowable value of this intensity for vascular applications is nearly an order of magnitude greater than that allowed for fetal examinations. This reflects the fact that the signal from flowing blood is small and that more power is needed to examine vessels lying deep in soft tissues which attenuate the ultrasonic energy.

Table 13.2 Maximum acoustic output of a modern scanner in transabdominal and transvaginal gray-scale imaging

Acoustic output	*Transabdominal*	*Transvaginal*	*FDA fetal*
I_{SPTA} (mW/cm^2)	10	5	180
I_{SPPA} (W/cm^2)	250	150	350
I_M (W/cm^2)	400	200	550

TRANSABDOMINAL AND TRANSVAGINAL GRAY-SCALE IMAGING

In transvaginal scanning, the transducer is placed close to the fetus. Some, therefore, have expressed concern that, if the same acoustic output is used in the transvaginal as in the transabdominal technique, the fetus would be insonated with more energy.

In general, transvaginal examinations are performed with higher frequency and the increased absorption by tissues reduces the anticipated fetal exposure.

Consider, for example, a transabdominal examination performed at a frequency of 3.5 MHz and that the I_{SPTA} is the maximum allowable 180 mW/cm^2. The National Council for Radiation Protection is currently considering the adoption of a model for the first trimester obstetrical examinations[2] in which the attenuation between the transducer and the fetus is independent of distance and linearly dependent on frequency, the value being given by the relationship 1 dB/MHz. Using this model, at 3.5 MHz, the attenuation between the transducer and the fetus is 3.5 dB and an intensity of 80 mW/cm^2 reaches the fetus.

Transvaginal examinations are usually performed at a frequency of 7.5 MHz. As yet, no model has been proposed for this examination. Because less intervening tissues are involved, a model in which the attenuation is 0.5 dB/MHz would seem reasonable. Using this model, the attenuation at 7.5 MHz is 3.75 dB, and the intensity which reaches the fetus is 76 mW/cm^2. The *in situ* intensity is, therefore, the same in the two techniques.

An alternative point of view which has been put forward is that a transvaginal examination should allow lower fetal exposure. Because the transducer is closer to the fetus, it is an efficient receiver of the signal reflected by the fetus. Therefore, for the same degree of visualization, the fetus needs to reflect less signal and does not have to be insonated with the same *in situ* intensity. Once differences in frequency and tissue distances are taken into account, reciprocity considerations similar to those described in the previous paragraphs indicate that the two factors balance out and the same *in situ* intensity strength is needed in both techniques.

Table 13.3 Minimum and maximum acoustic output in pulsed Doppler mode in transabdominal and transvaginal examinations

Acoustic output	*Transabdominal*	*Transvaginal*	*FDA fetal*
I_{SPTA} (mW/cm^2)			
minimum	150	50	180
maximum	850	150	
I_{SPPA} (W/cm^2)			
minimum	50	10	350
maximum	150	40	
I_M (W/cm^2)			
minimum	50	10	550
maximum	200	50	

Other factors actually govern the acoustic output used in the two approaches. The attenuation in the transvaginal technique is relatively constant in that the internal anatomical dimensions are not a strong function of the size of the patient. The transabdominal technique must cope with the examination of small as well as large patients and, for this reason, the equipment is designed to allow the transmission of a higher output at maximum output setting.

Table 13.2 lists the maximum acoustic output of a scanner used for transabdominal and transvaginal gray-scale imaging. As shown, maximum I_{SPTA} intensity in the transvaginal examination is half that available for transabdominal investigations. The intensity in both modes is an order of magnitude less than the FDA limit. Transabdominal or transvaginal gray-scale imaging is, therefore, unlikely to cause any significant thermal bioeffect.

Table 13.3 also illustrates that I_{SPPA} and I_M in transvaginal scanning are about half those used for transabdominal examinations. In all instances, the intensities are below the FDA limit. The FDA values incorporate a significant safety factor in that they are much less than the values at which any cavitational activity has been observed in water at the frequencies used in the two techniques. The possibility of any cavitational activity in transabdominal and transvaginal gray-scale imaging is, therefore, most unlikely.

PULSED DOPPLER EXPOSURE

Pulsed Doppler techniques are now well established in transabdominal and transvaginal applications. In this examination, in contrast to gray-scale imaging where the ultrasonic beam is scanned, the beam is kept stationary to receive the signal emanating from a fixed location. The pulse repetition rate is also increased to a value as high as is practical to allow measurement of high velocity flow. The result is that the I_{SPTA} is significantly higher than that used in gray-scale imaging and indeed approaches the limits quoted in the FDA document.

In the pulsed Doppler mode, significantly longer pulses are used as compared to those employed in gray-scale imaging. This is required to narrow the spectrum of the transmitted pulse and to allow calculation of the various Doppler parameters of the reflected signal. To keep the I_{SPTA} to within allowable limits, the equipment reduces the amplitude of the pulse, i.e reduces the I_{SPPA} and I_M.

The minimum and maximum acoustic outputs in the pulsed Doppler mode used for transabdominal and transvaginal examinations are shown in Table 13.3. As with gray-scale imaging, a control is provided over the transmit power for the pulsed Doppler examination. Transvaginal examinations are naturally limited to fetal examinations. For this reason, the acoustic output of transvaginal transducers at the high-output setting is limited to the FDA permissible output for fetal examinations. There is no natural distinction in transabdominal examination between a peripheral vascular and a fetal examination. For this reason, at the low power setting, the acoustic output is kept under the FDA limit so that transabdominal techniques may be used to examine the fetus. At the high power setting, the output is limited to the FDA figure stipulated for peripheral vascular applications. As mentioned, it is significantly higher than the fetal limit. User education and vigilance during the examination are necessary to ensure that transabdominal pulsed Doppler investigations are performed to conform with the FDA guidelines. For reasons discussed previously, the I_{SPPA} and I_M are less than those used in gray-scale imaging.

COLOR DOPPLER EXPOSURE

Pulses ranging from short ones as used in gray-scale imaging to longer ones as used in pulsed Doppler are used in color Doppler imaging.

In pulsed Doppler, the beam is kept stationary while the repetition rate is increased to allow measurement of high flow. Thus the same line of sight is laid down between 1000 and 5000 times a second. In color Doppler, the line of sight is scanned. Because the strength of the received Doppler signal is low, the same line of sight is laid down, typically 16 times, before the beam is scanned to the next spatial position. This explains the lower frame rate of color Doppler imaging which is typically of the order of 10 frames/second. Generally, the beam is scanned by a small amount so that there is some beam overlap between adjacent lines of sight. This results in some overlap exposure by adjacent lines of sight. A beam overlap factor is used to take this into account and in color Doppler it is typically a factor of 3. Thus in color Doppler, the same point in tissue is insonified by 16 x 10 x 3, i.e. 480, lines of sight. The I_{SPTA} in color Doppler is, therefore, 2–10 times less than that in pulsed Doppler.

Table 13.4 Maximum acoustic output from a transabdominal linear-array transducer in pulsed and color Doppler mode

Acoustic output	*Pulsed Doppler*	*Color Doppler*
I_{SPTA} (mW/cm^2)	1350	500
I_{SPPA} (W/cm^2)	300	300
I_M (W/cm^2)	600	600

Table 13.4 shows the maximum values of acoustic output of a transabdominal 5 MHz linear array transducer in pulsed and color Doppler modes. The pulsed Doppler I_{SPTA} is close to the FDA limit, while in the color mode it is about one-third that in the pulsed Doppler mode. Because the same pulse is used for color and pulsed Doppler, the I_{SPPA} and I_M are the same in the two techniques.

As yet, manufacturers do not provide information on the output of equipment functioning in the transvaginal color mode. There is no reason to expect that the principles discussed above would not apply. Reference to Table 13.3 would suggest that the I_{SPTA} would be less than half the FDA fetal limit and unlikely to cause significant thermal bioeffects.

CONCLUSION

Analysis of acoustic output of current sonographic equipment indicates that transabdominal and transvaginal gray-scale imaging are not capable of producing any significant biological effect.

The spatial peak-temporal average intensity, I_{SPTA}, in transvaginal color Doppler has not yet been reported. Indications are that it is well below the FDA permissible maximum output and is unlikely to cause significant thermal biological effects.

The I_{SPTA} reaches the FDA limit in transabdominal and transvaginal pulsed Doppler designed for fetal investigations at maximum output. Prudent use of the technique is, therefore, indicated, i.e. use of minimum output and dwell time consistent with obtaining the required diagnostic information.

Equipment at power settings designed for transabdominal peripheral vascular examinations exceeds by nearly a factor of 10 the FDA permissible maximum output for fetal investigations. As there is no natural barrier to a change from a peripheral vascular to a fetal examination, vigilance must be exercised that transabdominal fetal pulsed Doppler examinations are performed at appropriate power output settings and in a prudent manner.

REFERENCES

1. Kossoff, G. (1990). Current opinion on the safety of diagnostic ultrasound. In Kurjak, A. (ed.) *Handbook of Ultrasound in Obstetrics and Gynecology*, Vol. 1, pp. 47–53. (Boca Raton, Florida: CRC Press)
2. Carson, P., Rubin, J.M. and Chiang, E.H. (1989). Fetal depth and ultrasound path lengths through overlying tissues. *Ultrasound Med. Biol.*, **15**, 629–39

Index